Contents

Fuzzy Controllers

To
my son Dmitry
and other students

Fuzzy Controllers

LEONID REZNIK
Victoria University of Technology, Melbourne, Australia

Newnes
An imprint of Butterworth-Heinemann
Linacre House, Jordan Hill, Oxford OX2 8DP
A division of Reed Educational and Professional Publishing Ltd

A member of the Reed Elsevier plc group

OXFORD BOSTON JOHANNESBURG
MELBOURNE NEW DELHI SINGAPORE

First published 1997

British Library Cataloguing in Publication Data
A catalogue record for this book is available from the British Library

ISBN 0 7506 3429 4

Library of Congress Cataloguing in Publication Data
A catalogue record for this book is available from the Library of Congress

Typeset by The Midlands Book Typesetting Company, Loughborough, Leicestershire, England
Printed in Great Britain by Biddles Ltd, Guildford and King's Lynn

Foreword

Leonid Reznik's *Fuzzy Controllers* is unlike any other book on fuzzy control. In its own highly informal, idiosyncractic and yet very effective way, it succeeds in providing the reader with a wealth of information about fuzzy controllers. It does so with a minimum of mathematics and a surfeit of examples, illustrations and insightful descriptions of practical applications.

To view *Fuzzy Controllers* in a proper perspective a bit of history is in order. When I wrote my paper on fuzzy sets in 1965, my expectation was that the theory of fuzzy sets would find its main applications in fields such as economics, biology, medicine, psychology and linguistics – fields in which the conventional, differential-equation-based approaches to systems analysis are lacking in effectiveness. The reason for ineffectiveness, as I saw it, is that in such fields the standard assumption that classes have sharply defined boundaries is not a good fit to reality. In this context, it is natural to generalise the concept of a set by introducing the concept of grade of membership or, equivalently, allowing the characteristic function of a set to take values intermediate between 0 and 1.

Since my background was in systems analysis, it did not take me long to realise that the theory of fuzzy sets is of substantial relevance to systems analysis and, especially, to control. This perception was articulated in my 1971 paper 'Toward a theory of fuzzy systems', and 1972 paper, 'A rationale for fuzzy control'.

The pivotal paper was my 1973 paper, 'Outline of a new approach to the analysis of complex systems and decision processes', in which the basic concepts and techniques that underlie most of the practical applications of fuzzy set theory (or fuzzy logic, as we call it today), were introduced. The concepts in question are those of linguistic variable, fuzzy if-then rule and fuzzy rule sets. These concepts serve as the point of departure for what I call the theory of *fuzzy information granulation*. This theory postulates that in the context of fuzzy logic there are three basic modes of generalisation of a theory, method or approach: (a) fuzzification, in which one or more crisp sets are replaced by fuzzy sets; (b) granulation, in which an object is partitioned into a collection of granules, with a granule being a clump of points (objects) drawn together by indistinguishability, similarity or functionality; and (c) fuzzy granulation, in which a crisp or fuzzy object is partioned into fuzzy granules. In effect, fuzzy information granulation (f-granulation) is a combination of fuzzification and granulation.

What has not been recognised to the extent that it should is that the successes of fuzzy logic involve not just fuzzification but, more importantly, fuzzy granulation. Furthermore, fuzzy logic is the only methodology which provides a machinery for fuzzy information granulation. As we alluded to already, the key concepts underlying this machinery are those of linguistic variable, fuzzy if-then rule and fuzzy rule sets. Basically, fuzzy rule sets or, equivalently, fuzzy graphs, serve to provide a way of approximating to a function or a relation by a disjunction of Cartesian products of values of linguistic variables.

Viewed against this backdrop, it is – in effect, though not by name – the machinery of fuzzy information granulation that is employed in fuzzy controllers to explain – with high expository skills – what fuzzy controllers are, how they are designed, and how they are used in real-world applications. One cannot but be greatly impressed by the profusion of examples, the up-to-datedness of information, lucidity of style and reader-friendliness of Leonid Reznik's exposition. His work should have strong appeal to anyone who is looking for a very informative and easy to understand introduction to fuzzy controllers and their role in the conception, design and deployment of intelligent systems.

An issue of key importance in the design of fuzzy controllers is that of induction of rules from input-output data and tuning of their parameters. In the past, this was done by trial and error. More recently, techniques drawn from neurocomputing and genetic computing have been employed for this purpose. In *Fuzzy Controllers*, these techniques are discussed briefly but with insight in the last chapters. In these chapters, the reader will also find a very useful discussion of fuzzy system design software tools, their capabilities and their applications.

In sum, this book is an unconventional and yet very informative, self-contained and reader-friendly introduction to the basics of fuzzy logic and its application to the design of fuzzy controllers. Leonid Reznik deserves high marks for his achievement.

Lotfi A. Zadeh
Berkeley, CA

PREFACE

Since fuzzy logic was introduced by Lotfi Zadeh in 1965, it has had many successful applications mostly in control. This 'fuzzy' boom has generated strong interest in this area together with a boom in studying and teaching of fuzzy theory and technology. Although a few books which can be used for teaching and learning are available on the market, what is still missed is an introductory textbook suitable for both under- and postgraduate students, as well as for a beginner.

The aim of the book is to teach a reader how to design a fuzzy controller and to share some experience in design and applications. It can be used as a textbook by both teachers and students. Being an introduction this book tends to explain things starting from basics roots and does not require any preliminary knowledge in fuzzy theory and technology. I wanted to make this book different from other books available on the market. My goals were:

- to write a textbook that is intelligible even to a non-specialist;
- to pay attention, first of all, to practical aspects of fuzzy controller design;
- to facilitate the learning and teaching process for both a student and a teacher.

The structure of the book includes a description of the theoretical fundamentals of fuzzy logic as well as study of practical aspects of fuzzy technology. Consideration of all topics is practically oriented. This means that all the chapters work on achieving the final goal: to give a reader the knowledge necessary to design a fuzzy control system. To become a real textbook which can be used for self-assessment and teaching, this book contains the list of problems, assignment topics and design projects.

The style of the book changes from a textbook at the beginning (when it discusses theoretical aspects of fuzzy control) to a handbook (when it describes software and hardware tools which can be used in a fuzzy controller design). The book is written (especially at the beginning) as a discussion between a teacher and students who come from various educational and practical backgrounds and are supposed to be interested in different aspects of fuzzy control theory and technology.

I wanted to avoid making this book dull and boring, so I have tried to apply ordinary (not scientific) language without losing a correctness of mathematical determinations. It was very hard sometimes. That's why a few chapters (especially Chapter 2) contain a number of mathematical definitions and other constructions. However, I tried to provide the reader with some explanation about what all this mathematical stuff meant.

Acknowledgements

Many people from different countries and continents have contributed generously to the creation of this book. Unfortunately, I am not able to name everybody who has made this undertaking happen. It does not mean I do not remember all of you. Your valuable advice, encouragement and support is very much appreciated. Let me use this small space to give those names which I cannot miss.

A number of individuals, organisations and commercial companies have granted permission to reprint or adapt some material. I gratefully acknowledge the permission given by the Institute of Electrical and Electronics Engineering, Inc.; Adaptive Logic Inc.; David K. Kahaner, Director of Asian Technology Information Program, Tokyo, Japan; INFORM GmbH, Aachen, Germany; Inform Software Corporation, Chicago, USA and Constantin von Altrock; CICS Automation, Newcastle, Australia and Sam Crisafulli.

I wish to thank Professor Lotfi Zadeh for the invention of fuzzy logic and inspiration of this work as well as for finding time to read the manuscript and write this great foreword.

I would like to express my gratitude to my family, my parents Carl and Ninel, my wife Olga and son Dmitry for their patience.

Finally I want to acknowledge great work of the editors and publishers in shaping the manuscript and its preparation to the publication.

Feedback

I shall appreciate receiving any comments regarding this book, especially from teachers and students. Please do not hesitate to contact me directly or through the publishers.

Leonid Reznik
Victoria University of Technology
P.O. Box 14428 MCMC Melbourne 8001 Australia

Email: Leon Reznik@vut.edu.au

INTRODUCTION

Could you teach us very quickly how to design a fuzzy controller?

Well, it is not that easy...

Just explain a general method of design, optimisation, and implementation.

I do not think there is a general method of designing and finding optimal parameters for a fuzzy controller, because any such values always depend on the specific process or object under control and the control objectives. This is particularly the case for a fuzzy controller where the design process is very subjective.

However, I understand you want to start a design immediately. As you know nothing about fuzzy control and fuzzy controllers, we will base our design just on our human experience. Let us consider a classical control problem. We control the boat movement which we should drive along the straight line from point A to point B. How will you design a controller to do it?

Well, firstly I must derive a mathematical model of the plant and then develop a mathematical model of a controller.

And how will you develop this model?

I will apply one of the design methods, e.g., pole placement design, and obtain a transfer function for the controller.

OK. Now suppose we do not know exactly the mathematical model of our plant. Moreover, we do not know any classical design method. What can we do then?

Nothing. You have to know some theory. Otherwise you cannot do anything.

Try to use your own experience. How did you drive a boat in your childhood?

I turned the rudder left or right depending on the position of a boat.

Great! So we can say that if the boat is situated exactly on the line we should not do anything, if the boat is situated to the left of the line we should turn the rudder to the right (let us call this direction positive), and if the boat is positioned to the right of the line we should turn the rudder to the left.

Now let us try to formulate this control law as a set of rules.

If deviation is *zero* **then** turn is *zero.*
If deviation is *positive* **then** turn is *negative.*
If deviation is *negative* **then** turn is *positive.*

We have formulated the set of the control rules which can be written as a table, which is a mathematical model of our controller.

No! A mathematical model is either an equation or a mathematical function, or something like that.

Not necessarily. The mathematical model of a controller should describe mapping of an input, in our case it is the difference between the boat's current position and the desired one, to an output – control signal, in our case – the rudder angle. By using this table we can find the output corresponding to each value of the input.

Are you sure this table can control the boat movement?

Of course, very roughly. To improve the control quality and make our controller more reactive, we need to increase the number of values describing each variable. Until now we have not distinguished the values of the deviation and turn, but just considered their signs. Now let us use small, medium and big values for both the deviation and the turn. Then we obtain the rules table:

Deviation	NB	NM	NS	Z	PS	PM	PB
Turn	PB	PM	PS	Z	NS	NM	NB

How can we use this table? It just contains some abbreviations which I do not understand.

Actually these abbreviations represent the labels of fuzzy values. These labels give the 'fuzzy' value of the distance: big (PB and NB), medium (PM and NM), small (PS and NS), and very small (Z). The prefixes P and N represent the side of the line, that our boat deviates to, right or left, positive or negative.

So we describe our controller with the help of the rules table. This is just a rules base controller. Where is the fuzzy logic hidden?

In the processing of these rules. The problem of controller design includes not just the compilation of the rules table, but also the use of the table to calculate a control output. So we should also say how to process these rules in order to get the output result, and this processing is based on the fuzzy theory methods examined in Section 3.2. This describes the process of producing a fuzzy output from fuzzy inputs, which in the fuzzy set theory is called an inference engine.

Fuzzy inputs and outputs? Usually a controller uses measurement results, doesn't it? Are they fuzzy?

Not exactly. In fuzzy control, the measurement results or process outputs are generally assumed to be crisp, and are produced by technical devices (sensors). On the other hand, controller outputs should also be crisp to control different technical devices (actuators). It means that a fuzzy controller needs an interface at both input and output sides.

Have you started considering a structure of a fuzzy controller.

Yes. The process of fuzzy reasoning or processing the fuzzy rules is described in Section 3.2. Section 3.3 describes how a typical fuzzy controller operates. The structure of a simple fuzzy controller is described in Section 3.4.

What is a typical fuzzy controller?

It is not easy to define. We look at a PID-like fuzzy controller structure (Section 3.5) which has proved to be very popular with different designers in many applications. However, we propose a very brief consideration of more complicated structures and problems in Sections 3.5 and 6. Section 3.6 explains how to improve stability and performance of a fuzzy controller.

Both stability and performance?

You are actually asking how to evaluate the quality of a fuzzy controller design. In this book we are applying the criteria traditionally used in control engineering, especially in practical engineering, where one evaluates the controller performance based on the system response obtained.

I see this book operates with some terms from control engineering like a PID-controller, for example.

Yes. Basically there are two approaches to a fuzzy controller design: an expert approach and a control engineering approach. In the first (historically it was proposed earlier), the fuzzy controller structure and parameters choice are assumed to be the responsibility of the experts. Consequently, design and performance of a fuzzy controller depend mainly on the knowledge and experience of the experts, or intuition and professional feeling of a designer. This dependence, which is considered far from systematic and reliable, is the flaw of this approach. Chapter 3 helps to construct a fuzzy controller based on this experience. However, this approach could assist in constructing a fuzzy model or an initial version of a fuzzy controller.

The second approach supposes an application of the knowledge of control engineering and a design of a fuzzy controller in some aspects similar to the conventional design with the parameter's choice, depending on the information of their influence on the controller performance. This approach is conducted mainly in Chapter 4. This chapter opens Part II of the book, which is devoted to the problems of a fuzzy controller design.

So how do you describe a design practice?

A fuzzy controller design, as any other design process, consists of the following main steps: an initial choice of the controller structure and parameters (synthesis of the controller), together with a controller examination (testing) and an evaluation of the parameters' influence on the controller performance (analysis of the controller); an adjustment or change of the parameters and the structure based on the analysis results. The first step is considered in Chapter 4, and the last one in Chapter 5.

I still do not understand how you present practical design aspects.

Chapter 4 starts with the description of the practical design problems (Section 4.1) and proposed solutions. One of the problems considered here will be referred to at different stages of the design process. The chapter includes a brief description of a fuzzy controller design procedure (Section 4.2) and discusses in greater detail a choice of main parameters: scaling factors

(Section 4.3), membership functions (Section 4.4), fuzzy rules (Section 4.5), and defuzzification methods (Section 4.6).

In the second design approach, and to a lesser degree in the first approach, the initial parameter choice may be followed by the parameter tuning (a self-organising fuzzy controller) and the plant model formulation or modification (an adaptive fuzzy controller) – see Section 5.1. The methods of adjusting the fuzzy controller parameters are considered in Chapter 5. Classical tuning of fuzzy controller scaling factors is considered in Section 5.2 while an application of artificial neural networks and genetic/evolutionary algorithms are looked at in Sections 5.3 and 5.4. A brief description of the neuro-fuzzy controller design methodology is given in Section 5.3. Note that this chapter contains a considerable number of practical design examples.

Does one really need to design a fuzzy controller 'by hand'? I mean these days the design process is computerised and a designer applies different design packages. What is the situation with fuzzy controller design?

Basically the same as in other areas. A very fast-growing information technology industry has already developed and released a few good design packages which can be successfully applied in different applications for a fuzzy controller design. Among them are: RT/Fuzzy Toolbox for MATRIXx™ by Integrated Systems Inc., Fuzzy Logic Toolbox for MATLAB™ by The MathWorks Inc., FIDE™ by Aptronix, fuzzyTECH™ by Inform, a number of products by Togai InfraLogic Inc., Fuzzy Systems Engineering Inc., HyperLogic, etc. Some of them are specific for a fuzzy technology, others are universal and include a special fuzzy design toolbox.

The availability of these products on the market as well as their price means we do not include any software tools for fuzzy controller design in the text. Certainly our main advice is to use one of these packages (see Chapter 6). We will consider the features of the packages and give some examples of their applications.

Chapter 7 describes a realisation and a hardware implementation of fuzzy controllers. It gives advice on how to construct real fuzzy controllers. This is the most important part of the area of fuzzy controller design and implementation. New chips and devices are now being developed and manufactured in large quantities and any particular device will become obsolete before this book is read. So this chapter is short with only general descriptions and recommendations. However, it does include a very brief description of some of the latest hardware design achievements and products available on the market.

A considerable number of examples of practical fuzzy control system design are incorporated in the text. The goal of these examples is not just to illustrate theoretical assertions but to plunge the reader into the design environment.

Why have you started with Chapter 3?

This is because I have a problem with Chapter 2. While I did promise to avoid mathematical definitions and constructions, in order to have a more or less comprehensive understanding of fuzzy control the reader needs to grasp some basic mathematical concepts. As a matter of compromise, Chapter 2 explains what a fuzzy set is and what the difference between a fuzzy set and a crisp set is (Section 2.1), and what operations can be performed on fuzzy sets (Section 2.2).

Other parts of Chapter 2 are very important for understanding how a fuzzy controller works. Section 2.3 describes linguistic variables and hedges like Very Large, More or Less, Small, etc., which are applied in a fuzzy rules formulation. Section 2.4 describes how a conventional part of a control system processes fuzzy variables (fuzzy algebra). Section 2.5 gives a brief mathematical description of fuzzy processing (fuzzy relations).

If you like you can omit Chapter 2. Actually you can jump around this book, omitting any part of it. Part I considers theoretical fundamentals of a fuzzy controller operation, Part II looks at practical problems of fuzzy controller design, and Part III can be used as a manual in fuzzy controller design learning and teaching.

Part III starts with a brief manual on fuzzy controller design which summarises a design process and gives the reader concise advice on different aspects of a fuzzy controller operation, design, implementation and debugging. This part also includes a number of problems which can be used in teaching and for self-assessment. It proposes some possible topics for student assignments and projects. The first assignment is an essay comparing fuzzy control and technology with conventional methods. This approach establishes a general understanding of the place of fuzzy technology in modern science and industry. The combination of problem solving, essay writing and project design allows a teacher not just to cover mathematical fundamentals of fuzzy set theory, but to improve the theoretical and practical skills of students. The set of problems and design projects covers the fundamentals of fuzzy set theory as well as applications of fuzzy technology in different areas. So they could be useful for students and specialists in various disciplines: engineering (electrical, computer, mechanical, aerospace), information technology, computer and mathematical sciences. Problems are devoted mainly to particular aspects of fuzzy theory and technology, while projects cover a wider area and usually require a complex solution.

HOW DOES IT WORK? OR THE THEORY OF FUZZY CONTROL

1 FUZZY SETS, LOGIC AND CONTROL

1.1 Why do we need this new theory, what are the advantages of fuzzy control?

I have heard different terms: fuzzy set theory, fuzzy logic theory, fuzzy sets and logic. What is the difference? Do they denote different studies?

Usually people mean just one theory which they refer to under various names. This theory is rather young, which is why different people use different names even in English. In other languages, variations are much higher because of translations. However, you should carefully consider any particular case. Last year, some similar but different theories appeared, e.g. the theory of rough sets. The authors of these theories used some different axioms.

Different theories seem to appear. Why should we spend our time learning about this one? Is it worth studying?

Toshiro Terano pointed out that three conditions were necessary for a new theory [Ter94]:

- a societal need;
- a new methodology (both ideas and techniques);
- an attractiveness to researchers.

Let us consider these conditions. The aim of science and technology is to make our life easier. I understand that this is a controversial question, but do you agree that our life has become happier and simpler? Different people will give different answers. However, it is obvious that modern life includes large and complex organisations and sophisticated technical devices that need to be controlled. This requires the construction of mathematical models. Because these models are rather complicated and include some vagueness, it is hard to use classical mathematics to process these models. On the other hand, our brains possesses some special characteristics that enable it to learn and reason in a vague and uncertain environment. What to do?

How are you going to park a car?

Table 1.1 Limitations of conventional controllers

- Plant nonlinearity. The efficient linear models of the process or the object under control are too restrictive. Nonlinear models are computationally intensive and have complex stability problems.
- Plant uncertainty. A plant does not have accurate models due to uncertainty and lack of perfect knowledge.
- Multivariables, multiloops and environment constraints. Multivariate and multiloop systems have complex constraints and dependencies.
- Uncertainty in measurements. Uncertain measurements do not necessarily have stochastic noise models.
- Temporal behaviour. Plants, controllers, environments and their constraints vary with time. Moreover, time delays are difficult to model.

Maybe the answer is to try to model a human brain mathematically?

Right! A theory of creating and processing models, similar to those used by a human brain, was sought – Lotfi Zadeh proposed such a theory in 1965. The development of technology has computerised our life and strengthened the problem of man–machine interaction. Here I mean the man–machine interaction in a wide sense, not just as an interface but as a problem of establishing a harmony in communication between a computer and a human being on the levels of cooperative thinking, logic, language. We have a computer, operating according to Boolean logic with numerical mathematical models constructed by application researchers, and users who operate with another sort of logic and language including a high degree of ambiguity or fuzziness. Fuzzy sets theory aims to bridge this gap. It can be extremely useful not just in engineering and technological sciences but in social sciences, eliminating the difference in the approaches between natural and social sciences.

Table 1.2 Benefits of fuzzy controllers

- Fuzzy controllers are more robust than PID controllers because they can cover a much wider range of operating conditions than PID can, and can operate with noise and disturbances of different natures.
- Developing a fuzzy controller is cheaper than developing a model-based or other controller to do the same thing.
- Fuzzy controllers are customisable, since it is easier to understand and modify their rules, which not only use a human operator's strategy but also are expressed in natural linguistic terms.
- It is easy to learn how fuzzy controllers operate and how to design and apply them to a concrete application.

In the last two decades, the fuzzy sets theory has established itself as a new methodology for dealing with any sort of ambiguity and uncertainty. An underlying philosophy of the theory is a mathematical framework where imprecise conceptual phenomena in modelling and decision making may be precisely and rigorously studied. It lets mathematical models describe rather 'unmodelled' situations and finds solutions of 'unsolvable' problems. The theory includes a new mathematical apparatus and computer-realisable models.

The current number of researchers in this field and research societies mushrooming around the world show the attractiveness of this theory for both theoretical and practical researchers.

1.2 Where does fuzzy logic come from?

Fuzzy logic was introduced by Professor Lotfi Zadeh in 1965. Not in the least degree trying to undermine his achievements, this theory has its roots in the previous history of science, particularly in logic science. Although logic as a branch of Western science had been developing as binary logic, there were some famous paradoxes that could not be solved by binary logic. These paradoxes are as follows:

Falakros. Pluck a hair from a man's head and he does not suddenly become bald. Pull out another, and a third, and a fourth, and he still is not bald. Keep plucking and eventually the wincing man will have no hair at all on his head, yet he is not bald.

The Paradox of the Millet Seeds. Drop a millet seed on the ground and it makes no sound. But why is that dropping a bushel of millet seeds make a sound, since it contains only millet seeds? (After Zeno the Eleatic.)

Theseus' Ship. When Theseus returned from slaying the Minotaur, says Plutarch, the Athenians preserved his ship, and as planks rotted, they replaced them with new ones. When the first plank was replaced, everyone agreed it was still the same ship. Adding a second plank made no difference either. At some point the Athenians may have replaced every plank in the ship. Was it a different ship? At what point did it become one?

Wang's Paradox. If a number x is small, then $x + 1$ is also small. If $x + 1$ is small, then $x + 1 + 1$ is also small. Therefore five trillion is a small number and so is infinity. (After mathematician Hao Wang.)

Woodger's Paradox. An animal can belong to only one taxonomic family. Therefore, at many points in evolution a child must have belonged to a completely different family from its parents. But genetically, this feat is basically impossible. (After biologist John Woodger.)

Scientists tried to solve these obvious contradictions. Below [McNeil94] is a summary and some remarks on the problem of multivalued logic.

Plato (427–347? BC) saw degrees of truth everywhere and recoiled from them. 'No chair is perfect, it is only a chair to a certain degree.'

Charles Sanders Peirce (1839–1914) laughed at the 'sheep and goat separators' who split the world into true and false. 'All that exists is continuous and such continuums govern knowledge.'

Bertrand Russell (1872–1970) 'Both vagueness and precision are features of language, not reality. Vagueness clearly is a matter of degree.'

Jan Lukasiewicz (1878–1956) proposed a formal model of vagueness, a logic 'based on more values than TRUE or FALSE'. 1 stands for TRUE, 0 stands for FALSE, 1/2 stands for possible. Actually the three-valued logic by Lukasiewicz stayed just one step away from the multivalued fuzzy logic by Zadeh and can be considered as its closest relative.
Max Black (1909–89) proposed a degree as a measure of vagueness.

Albert Einstein (1879–1955): 'So far as the laws of mathematics refer to reality, they are not certain. And so far as they are certain, they do not refer to reality.'

Lotfi Zadeh (1923–) introduced fuzzy sets and logic theory. 'As the complexity of a system increases, our ability to make precise and significant statements about its behaviour diminishes until a threshold is reached beyond which precision and significance (or relevance) become almost mutually exclusive characteristics... A corollary principle may be stated succinctly as, 'The closer one looks at a real-world problem, the fuzzier becomes its solution.'

Table 1.3 describes the modern history of fuzzy logic after its invention by Zadeh in 1965. It is uncomprehensive and includes just some events but hopefully can be used for illustration of the fuzzy logic development.

Table 1.3 BRIEF HISTORY OF FUZZY TECHNOLOGY

Year	Event
1965	Concept of fuzzy sets theory by Lotfi Zadeh (USA)
1972	First working group on fuzzy systems in Japan by Toshiro Terano
1973	Paper about fuzzy algorithms by Zadeh (USA)
1974	Steam engine control by Ebrahim Mamdani (UK)
1977	First fuzzy expert system for loan applicant evaluation by Hans Zimmermann (Germany)
1980	Cement kiln control by F. – L. Smidth & Co. – Lauritz P. Holmblad (Denmark) – the first permanent industrial application
	Fuzzy logic chess and backgammon program – Hans Berliner (USA)
1984	Water treatment (chemical injection) control (Japan)
	Subway Sendai Transportation system control (Japan)
1985	First fuzzy chip developed by Masaki Togai and Hiroyuke Watanabe in Bell Labs (USA)
1986	Fuzzy expert system for diagnosing illnesses in Omron (Japan)
1987	Container crank control
	Tunnel excavation
	Soldering robot
	Automated aircraft vehicle landing
	Second IFSA Conference in Tokyo
	Togai InfraLogic Inc. – first fuzzy company in Irvine (USA)
1988	Kiln control by Yokogawa
	First dedicated fuzzy controller sold – Omron (Japan)
1989	Creation of Laboratory for International Fuzzy Engineering Research (LIFE) in Japan
1990	Fuzzy TV set by Sony (Japan)
	Fuzzy electronic eye by Fujitsu (Japan)
	Fuzzy Logic Systems Institute (FLSI) by Takeshi Yamakawa (Japan)
	Intelligent Systems Control Laboratory in Siemens (Germany)
1991	Fuzzy AI Promotion Centre (Japan)
	Educational kit by Motorola (USA)
After 1992	Too many events, inventions and projects to mention

1.3 What are the main areas of fuzzy logic applications?

Why did you stop in 1991? What are modern projects and inventions?

After 1991 fuzzy technology came out of scientific laboratories and became an industrial tool. Table 1.4 includes just a small number of successful projects and is intended to demonstrate a huge diversity of possible applications. On the other hand, Table 1.6 [Ter94] presents the current and future research topics being considered by Japanese researchers and engineers. One can see this table represents a good combination of various technical and social systems. It promises an interesting and fruitful future.

Table 1.4

- Automatic control of dam gates for hydroelectric power plants (Tokyo Electric Power.)
- Simplified control of robots (Hirota, Fuji Electric, Toshiba, Omron)
- Camera-aiming for the telecast of sporting events (Omron)
- Efficient and stable control of car engines (Nissan)
- Cruise-control for automobiles (Nissan, Subaru)
- Substitution of an expert for the assessment of stock exchange activities (Yamaichi, Hitachi)
- Optimised planning of bus timetables (Toshiba, Nippon-System, Keihan-Express)
- Archiving system for documents (Mitsubishi Elec.)
- Prediction system for early recognition of earthquakes (Seismology Bureau of Metrology, Japan)
- Medicine technology: cancer diagnosis (Kawasaki Medical School)
- Recognition of motives in pictures with video cameras (Canon, Minolta)
- Automatic motor-control for vacuum cleaners with a recognition of a surface condition and a degree of soiling (Matsushita)
- Back-light control for camcorders (Sanyo)

Tables 1.3 and 1.4 contain many very complex applications. Is it very difficult to understand how fuzzy logic works and how it is applied in these projects?

Not really. Firstly, fuzzy logic and fuzzy control feature a relative simplification of a control methodology description. This allows the application of a natural 'human' language to describe the problems and their fuzzy solutions. It fills a gap between modern scientific and technological devices and a simple 'man on a street'. Secondly, during the 'fuzzy' revolution, the fuzzy technology was introduced not only to a world of complex industrial projects, but to simple everyday home appliances (see Fig. 1.1).

Panasonic®/National® Fuzzy Logic

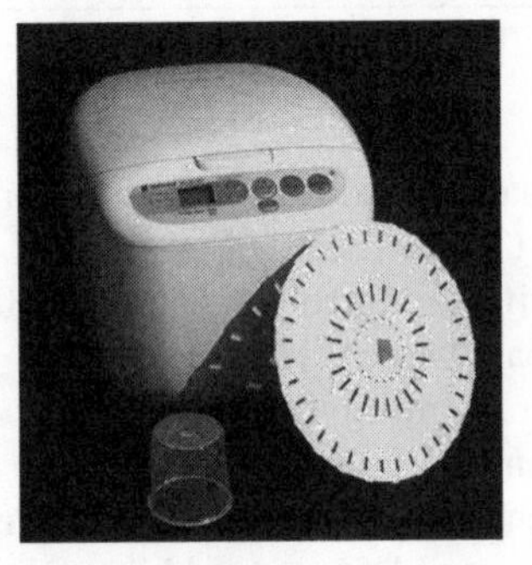

Rice cooker
Fuzzy logic controls the cooking process, self adjusting for rice and water conditions

National® Deluxe Electric Fuzzy Logic

Thermo pot
This unit represents the best technology available in producing clean boiled water on demand for making tea. It is fuzzy logic computer controlled

Fig. 1.1 *Two of the numerous fuzzy control home appliances*

I see the largest number of applications, included in the tables, are control applications. And most of the inventions are from Japan.

Table 1.5 Fuzzy controllers applications

Consumer products:
- washing machines
- microwave ovens
- rice cookers
- vacuum cleaners
- camcoders
- TVs and VCRs
- thermal rugs
- word translators

Systems:
- elevators
- train
- cranes
- automotive (engines, transmissions, brakes)
- traffic control

Software:
- medical diagnosis
- securities
- data compression

Absolutely right! These are two main paradoxes in the history of fuzzy logic. Although the fathers of fuzzy logic initially expected its main applications in large organisational system design and social sciences, most of real applications have been developed in engineering system control. And though the theory of fuzzy logic was born in the USA, the most rapid development in technology and applications has been in Japan. For example, one of the industrial pioneers in this area, OMRON, began to study fuzzy theory and technology seriously in 1984. Since then it has developed many kinds of fuzzy control-based products and has been granted many patents (more than 1000 in Japan and over 40 in the USA).

Fuzzy logic techniques have been adopted much more gradually in the USA and Europe. But the success of the Japanese developers is making companies like General Electric, General Motors, Hewlett-Packard, Rockwell and others take note. Starting in the 1990s, they have begun to apply it in their internal system development. For example, in the early 1990s, US companies started using fuzzy logic in the aerospace industry for applications such as rotor transmission, servo control, missile warning,

Fig. 1.2 *1996 Saturn SL1 and Mitsubishi Galant ES and LS are equipped with fuzzy controlled automatic transmission*

automated manufacturing and navigation systems. Meanwhile, General Motors has successfully incorporated fuzzy logic control into a product as widely used as the automatic transmission downshift mechanism of the Saturn™ (see Fig.1.2). However, Mitsubishi Galant™ S is offered with a standard five-speed overdrive manual gearbox and electronically controlled four-speed overdrive automatic transmission with 'fuzzy logic' shift control. Employing fuzzy logic, the electronic control unit of the transmission calibrates gear shifts and considers such inputs as a vehicle speed, throttle position and brake application, to determine whether the vehicle is going uphill or downhill and how twisty the road is before executing a shift.

Concluding this chapter, let me present two examples of successful fuzzy control applications. The information included here is taken mainly from the reports by David K. Kahaner [Kah95]. The first control system manages a very 'bad' plant which is unstable, nonlinear and subject to large disturbances, while the second one is developed to control the object with 'good' model behaviour.

Table 1.6

Machine systems	Human-based systems	Human machine systems
Picture/voice recognition	Human reliability model	Medical diagnosis
Chinese character recognition	Cognitive psychology	Inspection data processing
Natural language understanding	Thinking/behaviour models	Transfusion consultation
Intelligent robots	Sensory investigations	Expert systems
Crop recognition	Public awareness analysis	CAI
Process control	Risk management	CAD
Production management	Environmental assessment	Optimisation planning
Car/train operation	Human relations structures	Personnel management
Safety/maintenance systems	Demand trend models	Development planning
Breakdown diagnosis	Energy analysis	Equipment diagnostics
Electrical power system operation	Market selection models	Quality evaluation
Fuzzy controller	Category analysis	Insurance system
Home electrical	Social psychology	Human interface
appliance control		Management decision-making
Automatic operation		Multipurpose decision-making
		Knowledge bases
		Databases

Example 1.1 Fuzzy controller in an intelligent unmanned helicopter

One of the newest and most challenging technological applications is of the fuzzy logic in the control system of a helicopter, which is fundamentally unstable and with highly nonlinear dynamics. The helicopter Yamaha R-50 is a scaled down (3.6 metres head to tail) real helicopter with all the machinery for flying, plus all the control gears but minus the human accommodation. The engine, with the exhaust pipe, looks like one used in a Yamaha motorcycle. The development of R-50 was completed in 1987. The production number has increased each year from 70 (1990), to 100 (1991), 150 (1992), 200 (1993), and 250 in 1994. Its application is spreading to aerial photography, geographic, geological, oceanographic and coastal surveys, emergency rescue and fire control missions.

Fig. 1.3. *An unmanned voice-controlled helicopter [Schw92] (© IEEE 1992)*

Flight demonstration

The demonstration site is about the size of a baseball field. The helicopter must fly at a height of about 10 metres. The program consisted of four parts:

A. Command-based flight: a square path. The vehicle was to 'take off', do 'forward flight', '90 degrees righthand turn', 'forward flight' again, 'hover and turn left 90 degrees', 'fly backwards', 'hovering', 'fly sideways to the left', 'hover and then land'.

B. Command-based flight: zigzag. The vehicle was to 'fly forward', then 'fly to the right', 'fly to the left', then again 'fly to the right', 'fly to the left', and then 'hover'.

C. GPS (Global Position System) guided flight: from point A to point B while going through other two points, point #1 and point #2. It starts from point A, 'fly 20 metres forward to point #1', 'hover and turn right 90 degrees', 'fly 30 metres forward to point #2', 'hover and turn left 90 degrees' then 'fly 20 metres forward to point B'.

D. Image-guided flight: auto-landing. The vehicle was to 'take off', 'fly forward and search for the landing spot', 'tilt down the camera', 'hover and adjust position', 'adjust position as the vehicle lowers itself' and then 'land on the target spot'.

The flight programs proceeded smoothly without a hitch. The significance of this demonstration, however, is not what it looks like, but what goes into the control systems and what is involved in the new technology.

The reasons for applying fuzzy control. It has been over 50 years since the helicopter and conventional control technique were developed. The automatic control for the helicopter, however, has been limited to

- hovering control;
- maintaining the height after reaching a stable flight;
- change of route at intervals in accordance with the determined route.

Only partial automation has been accomplished. Most of the control has been manually operated.

Although unmanned helicopters, especially for military use, have been developed in every large country of the world, their control techniques have been confined to the remote control system using manual operation; there have been no papers written on the semi-remote control system (by language instruction employed) in this project because the conventional control has difficulty in achieving automatic operations.

Fuzzy control in view of the helicopter characteristics

A. Nonlinear behaviour: a helicopter has nonlinear characteristics. Conventional control methods use a linear theory only suitable for linear systems. In conventional control, the model is designed by a linear approximation around the equilibrium point of the helicopter, resulting in operational difficulties in states that deviate far from the equilibrium, and there is no guarantee for the performance. Fuzzy control is nonlinear and is thus suitable for the nonlinear system control.

B. Unstable system: the helicopter is intrinsically unstable, and there is a time delay between the input and output operations. It is thus difficult to achieve stability by the conventional feedback control which operates with further time lag.

C. Effect of the environment: a helicopter is very sensitive to the wind, for example. Exposure to a side wind leads to instability during hovering. As a pilot redirects the nose of the helicopter towards the wind, he attains stable hovering. Now there are no techniques associated with the conventional control method to deal with the change of the environment. However, fuzzy logic control can realise the pilot's stabilisation method by only adding the 'if–then' rules to accommodate with the environmental changes.

Example 1.2 Fuzzy subway control system (Sendai, Japan) [Yas85]

Another example of a major project is automatic train operation, which was installed in the Sendai Subway system, and has been in operation for approximately ten years. So this is a complete engineering system, not a prototype or experimental system.

A conventional train controller is based on proportional integral derivative (PID) control. The PID controller takes the error between the target speed and the actual train speed to control the train's motor and brake. A conventional control system requires a linearised system model, a desired state, and an error criterion, which is usually some function of the differences between desired and actual state. Given these, a PID control system can easily be implemented. This depends heavily upon analytic representations of the system and an error function as well as an assumption that the linearised model description does not deviate much from the real state. This method does not provide adequate control for systems with time-varying parameters or highly nonlinear systems, although often they can be tuned by incorporating detailed design information.

Humans can do still better because they can (on the basis of experience) evaluate the system objectives. Fuzzy predictive control is closer to a human operation. It attempts to evaluate not only the current system state, but also assesses the effect of a control command on the resulting state. As the performance indices are taken in human terms, good, very good, etc., this leads to the need to define the meaning of various linguistic performance indices.

The fuzzy predictive train controller applies a system model of the motor and brake to predict the next state of the speed, stopping point and running time (three-state vector) as inputs to a fuzzy controller. The fuzzy controller then selects the most likely control command based on the predicted state vector.

In many fuzzy control applications the model of the system

is unknown – it is for this reason that fuzzy control methods are usually chosen. But in this case the model is known precisely and is relatively simple because of the nature of the problem. In this case, fuzzy control is more concerned with the subjective measures for riding comfort, running time, and how close the train comes to a predesignated stopping point. Hence a distinction must be made between fuzzy control for unknown systems or nonlinear systems compared with fuzzy control with linguistic measures of the main objective.

The train operation is broadly classified into two control modes: (1) train speed regulation control, and (2) train stopping control. In the context of the train operation system, there were three key purposes:

- acceleration to a target speed;
- deciding and maintaining target speed;
- stopping accurately at a target position.

Also there are six performance indices:

- safety;
- accurate stopping;
- running time;
- energy consumption;
- comfort;
- traceability (maintaining target speed).

For the Sendai system, conventional PID automatic train operation control hardware was already being installed. Thus, incorporating predictive fuzzy control required only software changes to the onboard minicomputer. The system actually began commercial operation in 1987, and now operates through 17 stations over about 15 km. It is highly computerised, including:

- train tracking;
- route control;
- schedule management and adjustments;
- public address system (train approach, etc.);
- data transmission to/from train (trip pattern, etc.);
- depot in/out control;
- closed circuit TV in control centre when trains enter station;
- system monitoring;
- logging;
- system supervision (power, disaster prevention, etc.);
- business management (ticket sales, usage, etc.).

The use of fuzzy technology in automatic control of the Sendai Subway system exhibits several interesting facets:

- Fuzzy control is an efficient alternative to conventional control. Indeed, although automating train-driving operations can be viewed as simple, and conventional control can also be used, it does demonstrate that fuzzy control can do as good a job as conventional control.
- The Sendai Subway fuzzy control seems to have some advantages over a conventional control, such as riding comfort for passengers, savings in labour force and energy consumption, etc.
- From a technical viewpoint, it demonstrates that it is possible to use expert knowledge to design control laws and fuzzy theory as a means to translate natural language information into control strategies.

All your examples were from the engineering field. I am currently trying to persuade our business colleagues to utilise fuzzy control in an internal decision engine application and for fuzzy clustering. Could you provide us with an example of a financial application of fuzzy control?

Example 1.3 Financial evaluation and control

One of the most important applications of fuzzy logic was Yamaichi Fuzzy Fund. This is the premier financial application for trading systems. It handles 65 industries and a majority of the stocks listed on Nikkei Dow and consists of approximately 800 fuzzy rules. Rules are determined monthly by a group of experts and modified by senior business analysts as necessary. The system was tested for two years, and its performance in terms of the return and growth exceeds the Nikkei Average by over 20 per cent. While in testing, the system recommended 'sell' 18 days before Black Monday in 1987. The system went into commercial operations in 1988. All financial analysts including Western analysts will agree that the rules for trading are all 'fuzzy'.

Another example, which could be provided, is IBM which in January 1994 announced the first commercial fuzzy information systems application: a medical insurance fraud detection system. It demonstrates a high potential of fuzzy logic and control applications in 'non-traditional' areas.

2 BASIC MATHEMATICAL CONCEPTS OF FUZZY SETS

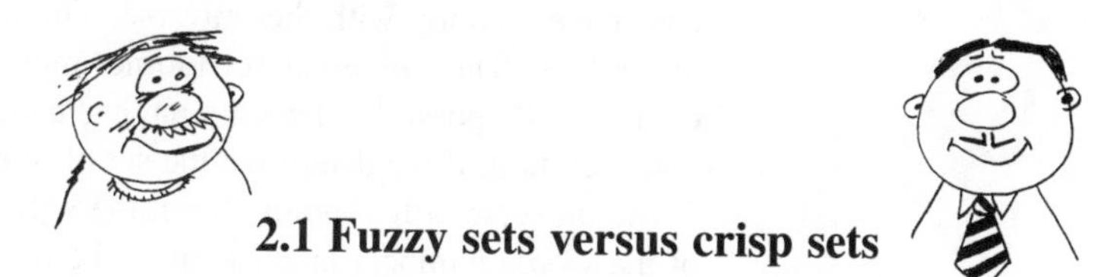

2.1 Fuzzy sets versus crisp sets

You said that fuzzy sets theory had many applications. Why do we call it fuzzy sets theory?

The theory of sets and the concept of a set itself constitute a foundation of modern mathematics. As far as one considers mathematical and simulation models of application problems, one deals with mathematics and the set theory at the base of mathematics. So if one changes this concept the whole building of modern science should be altered.

What is 'fuzzy' in a fuzzy set? How do you obtain a fuzzy set from a crisp one?

Let us remember how we determine a crisp set: It is a collection of objects of any kind, numbers, geometric points, chairs, pencils, etc. Usually the set is determined by naming all its members (the list method) or by specifying some well-defined properties satisfied by its members (the rule method). In the second method, a rule allows us to determine whether any particular element of the universal set belongs to the set under determination, or it does not. Similarly, one can name the elements of the universal set which do not belong to our set. Then all the other elements of the universal set compose the determined set. To indicate that an individual object x is a member of a set A, we write $x \in A$. Whenever x is not an element of a set A, we write $x \notin A$.

I understand that we can compose the set from some numbers, e.g. A= {3,4,5,6} by naming them. But how can we name, say, all positive numbers?

The list method can be used for finite sets only. When you determined your set, you actually used the rule method (rule: all members are positive numbers). And by this rule you define which elements of the universal set belong to your set.

Fine. I intuitively understand what the set is. But could you give a more mathematical definition, please?

I am sorry, I can't. As I have already said, the concept of the universal set is a basic concept which is difficult to determine. We can make analogy with the universe. The universe is everything around us. The universal set (sometimes called the universe) contains all possible elements having nature or property under consideration. If we determine the set of the numbers larger than 3, the universe will contain all numbers. If we determine the set of the *whole* numbers larger than 3, the universe will contain all *whole* numbers. If we consider the set of long pencils, the universe will contain the set of all pencils. The number of elements of the universal set is always larger than or equal to the number of elements of the set considered.

Then any set is a subset of the universal set, isn't it?

That's right! Any set A can be considered as a subset of the universe U and we can write $A \subset U$. There is another important set, the empty set $\varnothing$. While the universe contains everything, this set contains nothing, no elements.

You mentioned the set of long pencils. Can a set include not just numbers?

Yes, it can. The elements of the set can have any nature: we can have a set of points, a set of houses, a set of students, etc. You need to remember just that the nature of the elements of any particular set should coincide with the nature of the universe on which it is determined.

If we take the universal set as a base, then we can easily determine any particular set (or strictly speaking a subset of the universal set) by saying if any element of the universal set does or does not belong to this set. This process, by which individuals from the universal set X are determined to be either members or non-members of a set, is mapping the universal set into the determined set, and as mapping it can be defined by a function which is usually called a characteristic function. For a given set A this function assigns the value $\mu_A(x)$ to every $x \in X$. Guess how many values may this function have?

This function determines if the element of the universal set does or does not belong to this set A. Hence the function may have two values: TRUE or FALSE.

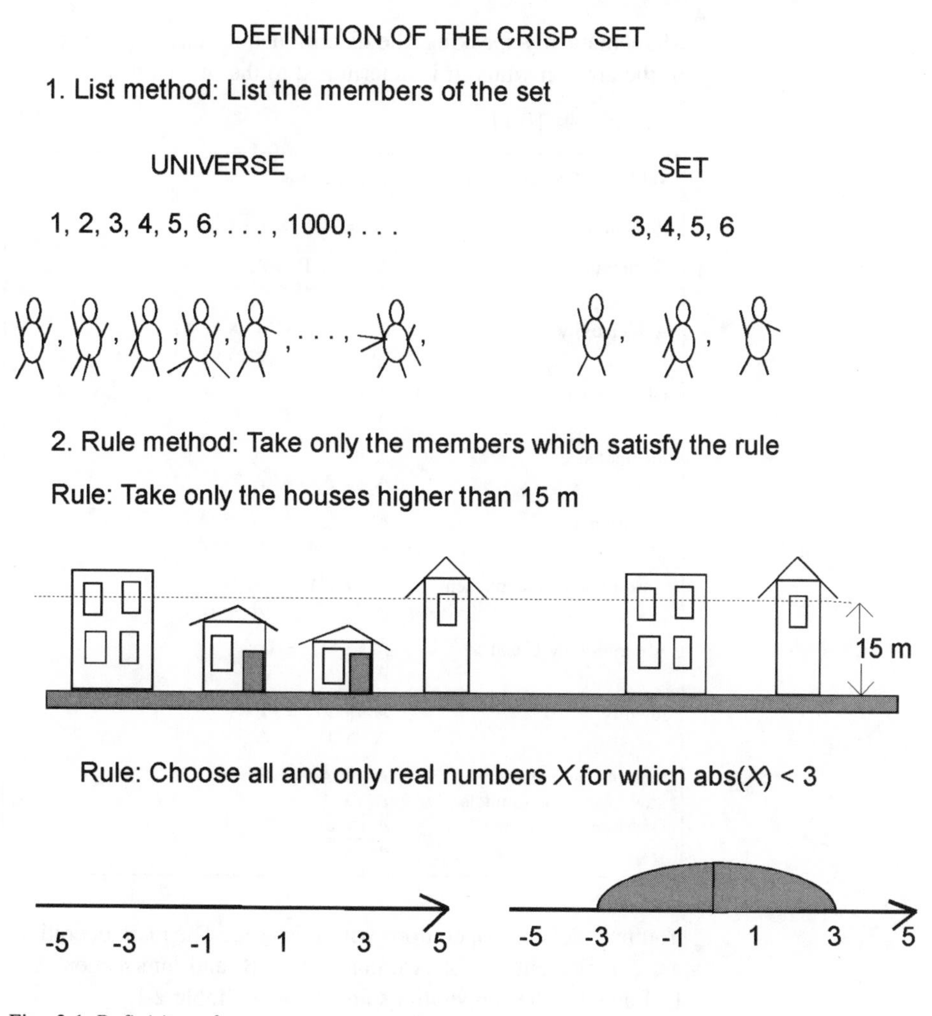

Fig. 2.1 *Definition of a set*

That's right. If we use Boolean function, we will have TRUE or FALSE, or in numbers, 1 or 0.

$$\mu_A(x) = \begin{cases} 1 \text{ if and only if } x \in A \\ 0 \text{ if and only if } x \notin A \end{cases}$$

Sometimes this function is called a discrimination function, because it discriminates the elements of the universal set which belong to the set from those which do not. Determining the discrimination function, we divide the universal set into two parts. So we have the certain discrimination and the certain border between these two parts.

Mathematically speaking, the discrimination function is mapping of the area on which it is determined to the two-elements set:

$$\mu_A : X \rightarrow \{0,1\} \quad (2.1)$$

Table 2.1 Properties of crisp set operations

Involution	$(\bar{A})' = A$
Commutativity	$A \cup B = B \cup A$ $A \cap B = B \cap A$
Associativity	$(A \cap B) \cap C = A \cap (B \cap C)$ $(A \cup B) \cup C = A \cup (B \cup C)$
Distributivity	$A \cup (B \cap C) = (A \cup B) \cap (A \cup C)$ $A \cap (B \cup C) = (A \cap B) \cup (A \cap C)$
Idempotence	$A \cup A = A$ $A \cap A = A$
Absorption	$A \cap (A \cup B) = A$ $A \cup (A \cap B) = A$
Absorption of complement	$A \cup (\bar{A} \cap B) = A \cup B$ $A \cap (\bar{A} \cup B) = A \cap B$
Absorption by U and ∅	$A \cup U = U$ $A \cap \varnothing = \varnothing$
Identity	$A \cup \varnothing = A$ $A \cap U = A$
Law of contradiction	$A \cap \bar{A} = \varnothing$
Law of excluded middle	$A \cup \bar{A} = U$
De Morgan's laws	$\overline{A \cap B} = \bar{A} \cup \bar{B}$ $\overline{A \cup B} = \bar{A} \cap \bar{B}$

You may do some operations with crisp sets. The main operations are complement $\bar{A}$ (or A'), union $A \cup B$, and intersection $A \cap B$. Laws for these operations are shown in Table 2.1.

I do not understand how to implement these operations.

Let us circle the universal set consisting of all points inside the bold line in Fig. 2.2.

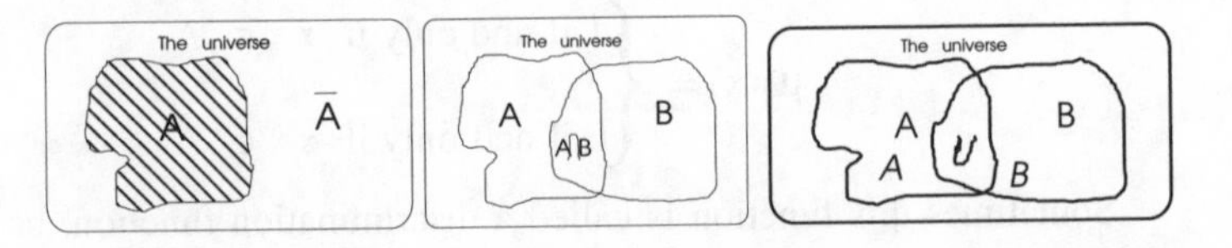

Fig. 2.2 *Operations on crisp sets*

This figure illustrates the operations of complement, intersection and union.

Another example: consider all the students in a class as the universal set. All students younger than 20 are as set A and all male students are set B. Then the complement of A or $\bar{A}$ contains all students older than 20 years, and complement B or B contains all female students.

The intersection of A and B includes all male students that are younger than 20. What about the union?

The union should include all male students and all female students younger than 20. It gives us all students except females older than 20. This statement can be obtained from De Morgan's law.

Is there a more mathematical definition of operations?

Let us use the characteristic function. Then the complement of A or $\bar{A}$ is determined by the characteristic function $\mu_{\bar{A}}(u) = 1 - \mu_A(u)$. The intersection $A \cap B$ can be determined with the characteristic function, for example

$$\mu_{A \cap B}(u) = \min(\mu_A(u), \mu_B(u)).$$

What do you mean saying 'for example', 'can be determined'? How can it be determined otherwise?

The choice of these operations is arbitrary. The only property the intersection operation has to satisfy is to return 1 if both arguments are 1, and to return 0 otherwise. Propose another formula to calculate the intersection characteristic function.

I think we can use $\mu_{A \wedge B}(u) = \mu_A(u) \times \mu_B(u)$.

Beautiful! Now what property should the union characteristic function satisfy?

It has to return 1 if one of the arguments is 1, and 0 otherwise.

Well done! The following formulae realises it.

$$\mu_{A \cup B}(u) = \max(\mu_A(u), \mu_B(u))$$

Now let us expand a set of possible values of the characteristic function from the two-element set to the continuum set (the set of all real numbers between 0 and 1). Let us suppose that the characteristic function may have any value between 0 and 1, not just 0 and 1. Then we can replace (2.1) with

$$\mu_A: X \rightarrow [0,1]$$

Mathematically it is very easy. But what does it mean in the real world?

It means that now we do not determine exactly whether any particular element of the universal set belongs or does not belong to the set A. We are saying that an element belongs to the set A with the determined membership degree. This degree is assessed by a number between 0 and 1.

I understand, that if for the element x its membership degree is 0 it does not belong to the set A, and if it is 1 it does. What will happen if it is 0.5?

It means we do not know whether this element of the universal set does or does not belong to this set A.

And if it is 0.8?

We still do not know exactly, but we think the possibility of belonging is higher.

Actually we fuzzify the border between two parts of the universal set. We expand the set of possible characteristic function values and get the fuzzy set. In fuzzy set theory the characteristic function is usually called the membership function.

But it means that the element can be simultaneously in both parts. How can it come?

If we consider Boolean logic where any variable can have just two values: TRUE or FALSE, it is impossible. A classic computer works according to the rules of this logic. However, any real life situation is much richer than this black and white model. A real object usually contains a lot of grades of a grey colour between black and white. So a real object can be more or less black, white, short, long, young, old, etc.

This approach makes the theory more complicated. Two possible values are replaced with a continuum set of values. The discrimination function is replaced with the membership function. Other properties of the sets should be changed as well.

From the computer point of view (if such a point exists) you are right. However, a human intellect has had experience of processing this information for ages. I would say that this model was more general than the previous one. It is obviously closer to the models, that a human being creates in his/her mind. This approach tends to decrease the gap between natural intellect models produced by human experts and artificial intelligence models used by computers.

OK. However, don't these models have to be followed by more complicated algorithms and procedures in solving application problems?

Not always. As we will see later, the trend to use 'accurate' models for describing real objects and processes often leads to the sophisticated method application. Using fuzzy models, one can significantly simplify models and corresponding procedures. Let us leave it for a while and come back and define the basic terms of fuzzy sets.

> Definition 2.1 Fuzzy set
>
> A fuzzy set A in the universal set U is a set of ordered pairs of a generic element u and its membership degree $\mu_A(u)$ as $A = \{ (u, \mu_A(u))/ u \in U\}$

I see the fuzzy set definition is similar to the crisp set one. To determine a crisp set we name all the elements of the universe which belong to the set A. In a fuzzy set we name all the elements of the universe and supplement to them a number between 0 and 1. This number demonstrates to what degree this generic element belongs to the defined fuzzy set. Actually, according to this definition we just add to every element a number, which constitutes the membership degree of this element.

Actually, a fuzzy set is given by its membership function. The value of this function determines if the element belongs to the fuzzy set and in what degree.

DEFINITION OF THE FUZZY SET

1. List method: List the members of the set

UNIVERSE

1, 2, 3, 4, 5, 6, . . . , 1000, . . .

FUZZY SET

1/0.1, 2/0.2, 3/0.9, 4/0.95, 5/0.9, 6/0.85, 7/0.5, . . .

/0.8 /0.3 /0.7, /0.9 /0.7 . . . , /0.1

2. Rule method: Take only the members which satisfy to the rule

Rule: Take only the high houses

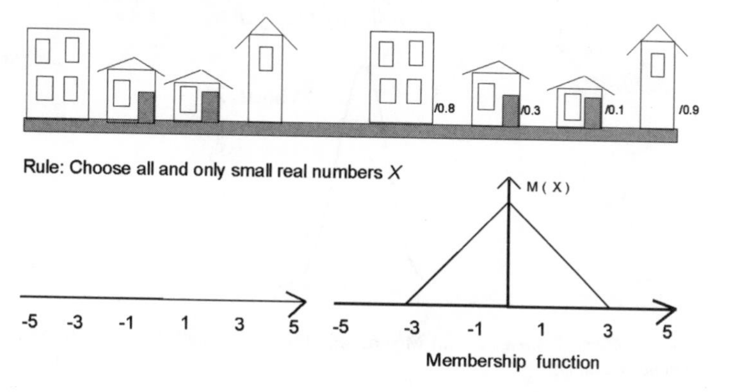

Fig. 2.3 *Definition of the fuzzy set*

Definition 2.2 Support

The support S(A) of a fuzzy set A is the crisp set of all the elements of the universal set for which membership function has non-zero value

$$S(A) = \{ u \in U / \mu_A(u) > 0 \}$$

I see. Only those elements should be considered that have any chance of becoming members of the fuzzy set.

Yes. Depending on the universal set, a fuzzy set can have a finite or an infinite number of elements. The support can be a finite set. It means that just a finite number of elements of the universal set have a non-zero membership degree.

Definition 2.3 Crossover point

The element of the universal set, for which the membership function has the value of 0.5, is called a crossover point.

The crossover element marks the point where the possibility of belonging becomes lower than the possibility of not belonging. Although an exact position of a crossover point is usually not very important, it characterises a shape of the membership function.

Example 2.1 *Consider a set of five pencils located in the box. Determine a fuzzy set of 'short pencils' A as*
A= {pencil1/0.2, pencil2/0.5, pencil3/1.0, pencil4/1.0, pencil5/0.9}

In this example pencil3 and pencil4 are exactly short, pencil5 is almost short, pencil2 is more or less short and pencil1 is almost exactly not short.

Example 2.2 *If the support is infinite we can consider membership functions in Fig. 2.4.*

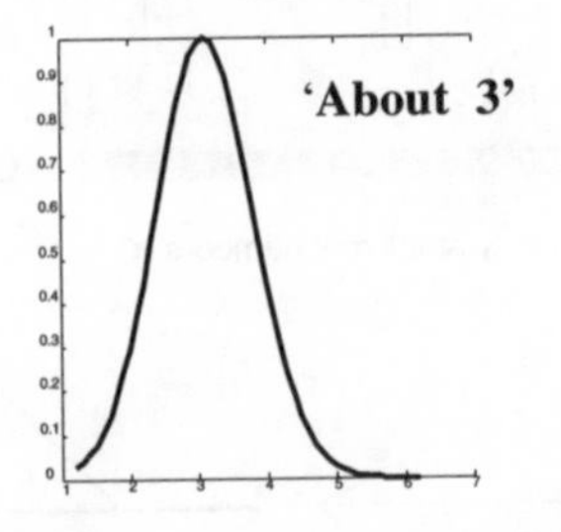

a) Membership function 'about 3'

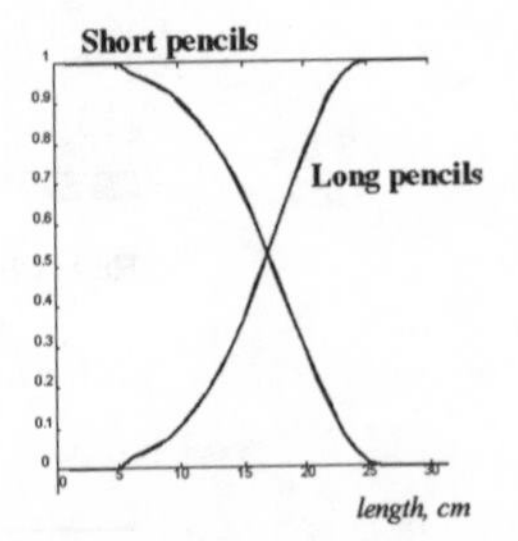

b) Membership functions for 'short' and 'long' pencils

Fig. 2.4 *Membership functions examples*

In this figure we have different shapes of membership functions.

You are right. The most often used functions are:

- piecewise linear (triangular and trapezoidal);
- quadratic;
- gaussian according to the formula $\mu(x) = \exp(-(x - \mu)/\sigma)^2)$;
- some special functions.

Which one is better to use?

Triangular or trapezoidal (piecewise linear) functions have proved to be more popular with fuzzy logic theoretics and practitioners rather than higher order based functions such as quadratic, cubic, etc. A possible reason for this is a simplicity of this function often allowing for the prediction and calculation of an output of the fuzzy system. Another reason is that the extra smoothness introduced by higher order fuzzy sets and demanding higher computational consumption is not strongly reflected in the output quality of a fuzzy model. However, the problem of the membership function choice has not yet been solved theoretically. Different researchers choose numerous shapes in various application problems.

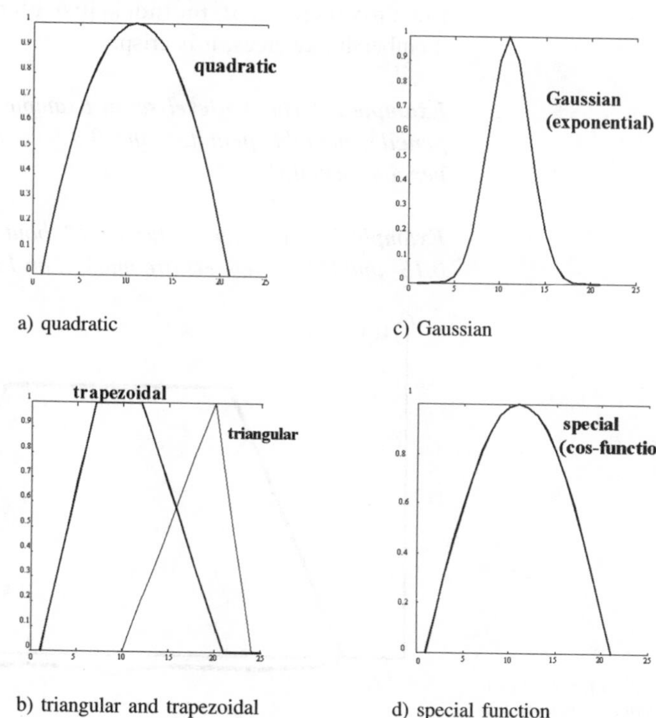

a) quadratic

c) Gaussian

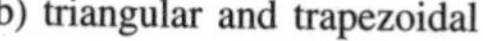

b) triangular and trapezoidal

d) special function

Fig. 2.5 *The most widely used membership function shapes*

A shape determines how many elements of the universal set have the membership degree larger than a given value between 0 and 1.

> Definition 2.4 Level set
>
> The set of elements that belong to the fuzzy set A at least to the degree α is called the α-level-set or α-cut-set
>
> $A_\alpha = \{ u \in U / \mu_A(u) > \alpha \}$

This definition gives us another way to determine a fuzzy set. As any membership function has all its values between 0 and 1, we can construct a membership function on the base of the level sets. And because a membership function determines a fuzzy set completely, any fuzzy set A can be considered as the union of all its level sets: $A = \cup_\alpha A_\alpha$, $\alpha \in [0,1]$.
By the way, what do you think, either an α-level-set is crisp or fuzzy?

As a subset of a fuzzy set, it is fuzzy.

I do not agree that it is a subset of a fuzzy set. This is another set, it contains the elements of the universe, so it is a subset of the universe. As it includes just elements without their membership degrees, it is crisp.

Example 2.3 *The 0.5-level-set in Example 2.1 contains {pencil2, pencil3, pencil4, pencil5}, the 0.8-level-set contains {pencil3, pencil4, pencil5}.*

Example 2.4 *Consider the fuzzy set 'about 3' of Example 2.2. The 0.15- and 0.45-level-sets are marked on Fig. 2.7.*

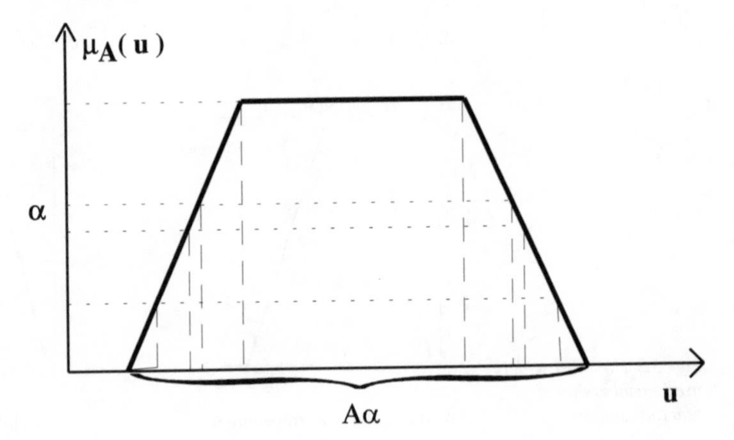

Fig. 2.6. *A fuzzy set as a collection of α-level sets*

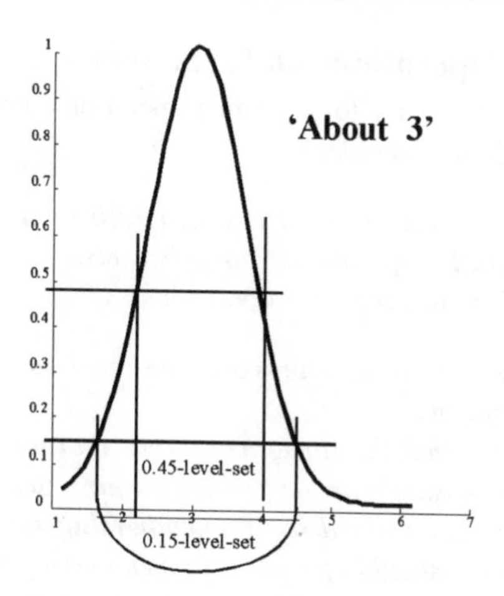

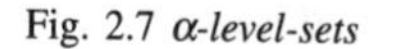
Fig. 2.7 *α-level-sets*

> Definition 2.5 Height of a fuzzy set
>
> **The height of a fuzzy set A, hgt(A) is given by a supremum of the membership function over all $u \in U$ $hgt(A) = \sup_U \mu_A(u)$**
>
> (Supremum in this definition means the highest possible (or almost possible) degree.)

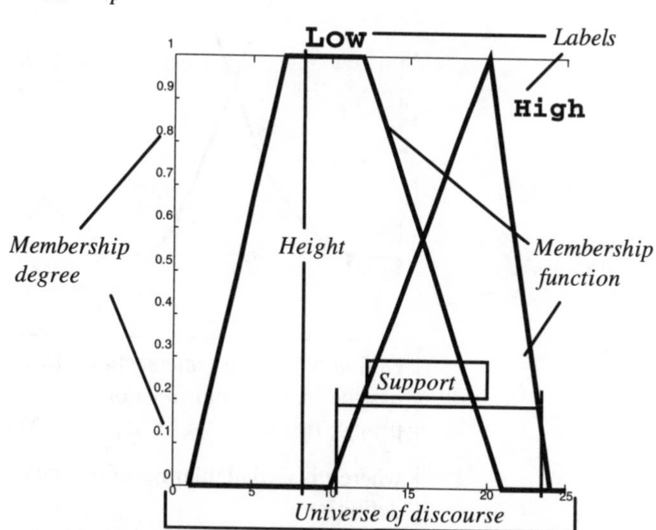

Fig. 2.8 *Main terms of membership functions*

2.2 Operations on fuzzy sets

Now let us try to determine operations with fuzzy sets. How do we do it? Any idea?

Because the fuzzy set is a generalisation of a crisp set, we can generalise operations with crisp sets.

How to generalise operations? What do you mean?

I do not know. The word 'generalisation' sounds nice and significant.
I guess that the same operations, that have been determined on crisp sets, can be determined on fuzzy sets. And because a fuzzy set is determined by its membership function, the operations actually should operate with membership functions.

Exactly! And main operations are complement, intersection (or conjunction) and union (or disjunction).

> Definition 2.6 The complement of a fuzzy set A has a membership function which is pointwise defined for all $u \in U$ by $\mu_A(u) = 1 - \mu_A(u)$.

This corresponds to the logical NOT operation. It is quite simple.

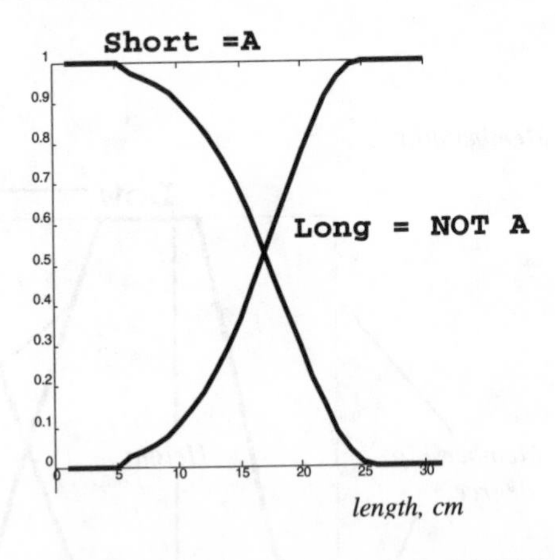

Fig. 2.9 *Complement operation on fuzzy sets*

> Definition 2.7 The intersection of two fuzzy sets $C = A \cap B$ has the membership function
> $\mu_{A \cap B}(u) = \mu_A(u) \; \widehat{t} \; \mu_B(u) \le \mathrm{Min}(\mu_A(u), \mu_B(u))$
> where t is a triangular norm (or measure) defined by the next definition.

Definition 2.8 The $\hat{t}$ norm
The triangular norm is a two-place function from [0,1] * [0,1] to [0,1], i.e. $\hat{t}$: [0,1] * [0,1] → [0,1] which is non-decreasing in each element $x \, \hat{t} \, w \leq y \, \hat{t} \, z$ if $x \leq y$, $w \leq z$ commutative, associative and satisfies boundary conditions x $\hat{t}$ 0 = 0, and x $\hat{t}$ 1 = x; for any x,y,z,w ∈ [0,1].

I do not understand this definition. How can we use it? How can we calculate the membership function of the intersection? It cannot be applied as an operation definition.
Why not?

A definition should explain how to make an operation. From this definition we do not know what to do with membership functions in order to obtain a result. Should we add them, multiply or make logical operations?

One can use different ways and various formulas to calculate the triangular norm. Some of them are given in Table 2.2.

Table 2.2 Some classes of fuzzy set unions and intersections

Reference	Fuzzy unions	Fuzzy Intersections	Range of parameter
	max (a,b)	min (a,b)	
	a + b – ab	ab	
Schweizer & Sklar [1961]	$1 - \max[0, (1-a)^{-p} + (1+b)^{-p} - 1)]^{1/p}$	$\max(0, a^{-p} + b^{-p} - 1)^{-1/p}$	$p \in (-\infty, \infty)$
Hamacher [1978]	$(a + b - (2-g)ab)/(1 - (1-g)ab)$	$ab/(g + (1-g)(a + b - ab))$	$g \in (0, \infty)$
Frank [1979]	$1 - \log_s [1 + (s^{1-a} - 1)(s^{1-b} - 1)/s - 1]$	$\log_s [1 + (s^a - 1)(s^b - 1)/s - 1]$	$s \in (0, \infty)$
Yager [1980]	$\min[1, (a^w + b^w)^{1/w}]$	$1 - \min[1, (1-a)^w + (1-b)^w)^{1/w}]$	$w \in (0, \infty)$
Dubois & Prade [1980]	$\dfrac{a + b - ab - \min(a,b,1-\alpha)}{\max(1-a, 1-b, \alpha)}$	$ab/\max(a,b,\alpha)$	$\alpha \in (0,1)$
Dombi [1982]	$\dfrac{1}{1 + [(1/a - 1)^{-\lambda} + (1/b - 1)^{-\lambda}]^{-1/\lambda}}$	$\dfrac{1}{1 + [(1/a - 1)^{\lambda} + (1/b - 1)^{\lambda}]^{1/\lambda}}$	$\lambda \in (0, \infty)$

Anyone can propose his or her own definition!?

You're right! It is a good exercise to introduce a new way to calculate the triangular norm. Try to write down a few.

Can we compare results of these definitions? Maybe different ways of calculation produce different results.

They definitely do (see Fig. 2.10). On this figure, two operations, min and product, were applied. And you see they have produced different results. For example, for the element u = 5, if we calculate the $\hat{t}$ norm operation as a product, the membership degree will be 0.15. However, if one calculates it as a minimum the degree will be 0.35. So, the difference is quite significant.

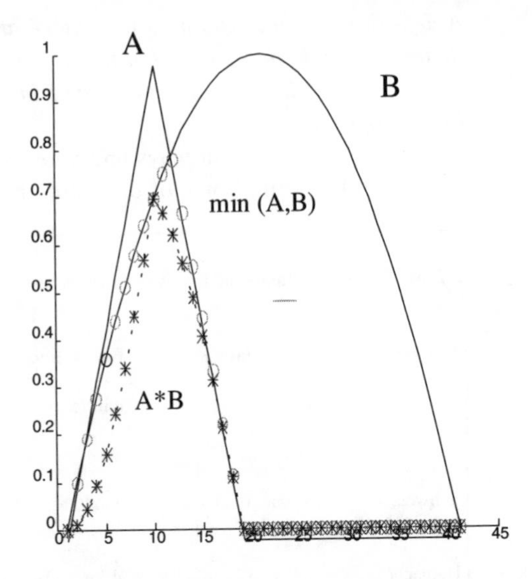

Fig. 2.10. *Two different ways for calculating t^{l}–norm operations*

You see yourself your definition is wrong because it can be realised by different ways which give different results.

Nothing is wrong here. You can use any definition if it satisfies the conditions of Definition 2.8. We can compare these different results, e.g. the usual $\hat{t}$norms are intersection and algebraic product with $x \cap y \geq x \times y$. Calculate the other results and compare them.

> Definition 2.9 The union (or disjunction) of two fuzzy sets $C = A \cup B$ has the membership function
> $\mu_{A \cup B}(u) = \mu_A(u) \;\; \hat{s} \;\; \mu_B(u) \;\geq\; \max(\mu_A(u),\; \mu_B(u))$
> where $\hat{s}$ is a triangular co–norm (measure or $\hat{s}$ norm) defined below.

Definition 2.10. The $\hat{s}$ norm
The triangular co-norm is a two-place function from [0,1] * [0,1] into [0,1], i.e. s: [0,1] * [0,1] → [0,1] which is non-decreasing in each element, commutative, associative and satisfies boundary conditions $x\ \hat{s}\ 0 = x$, and $x\ \hat{s}\ 1 = 1$; for $x \in [0,1]$.

As far as I see, the triangular norm and the co-norm are rather similar.

The definitions are similar and they have a determined relationship with each other. This relationship is given by the equivalent of De Morgan's law in set theory:

$$x\ \hat{s}\ y = 1 - (1 - x)\ \hat{t}\ (1 - y), \quad \text{for } x,y \in [0,1] \qquad (2.2)$$

Now, by how many ways can the triangular co-norm be calculated?

I would not be surprised if it had plenty of ways.

That's right! Some of them are given in Table 2.2.

Which way is the best to use?

I do not know the answer. Different researchers propose and apply different ways. Even computer packages include different opportunities, which can be chosen by a user. The most commonly used are intersection and algebraic product. Few experiments have been carried out as to the optimal triangular norm selection. The min (max) operators, for example, indicate no set interaction, since when one membership function is smaller (larger) with respect to the other, it has no influence on the resultant membership function.

Is it good or bad?

This robustness is an advantage if the accurate values of membership functions are not required. However, it makes the result less sensitive to changes in one of the arguments.

To illustrate the idea of different calculations [Drian93] uses the example by Van Nauta Lemke. First consider Fig. 2.11, representing a prison cell with two windows. Suppose we have two different problems to be solved:

1. How easy is it for a prisoner to escape from his prison cell?
2. How easy is it for sunlight to come into a prison cell?

Suppose window 1 is very difficult to escape from (the 'simplicity' is 0.1) and window 2 is easier (the simplicity is 0.3). Suppose also that window 1 is dirty and thick such that it is not so easy to penetrate it (0.6), while window 2 is clean (0.9).

The answer to problem 1 should be min (0.1, 0.3) = 0.1 if we do not take into account the fact that the person has to be fast or may get tired, etc.

In problem 2, the answer is $0.6 \times 0.9 = 0.54$, because the sunlight should penetrate both windows.

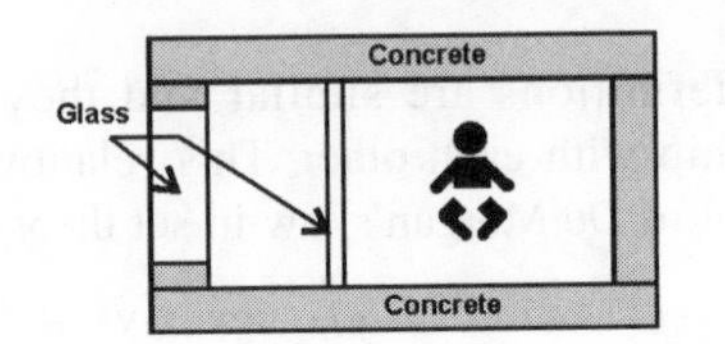

Fig. 2.11 *The prisoner problem.: How easy to escape? How easy to penetrate?*

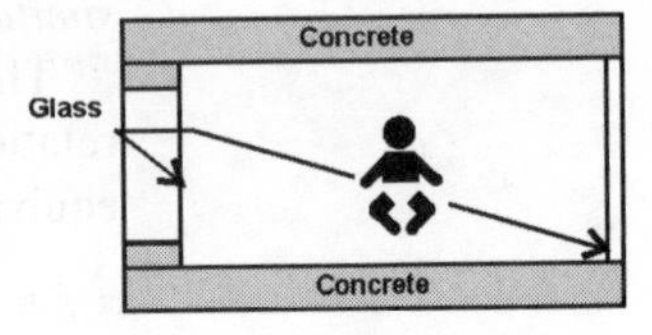

Fig. 2.12. *The prisoner problem (another configuration): How easy to escape? How easy to penetrate?*

In Fig. 2.12 the cell configuration has changed. What operation do we model here?

In Fig. 2.11 we model the operation AND (intersection) and in Fig. 2.12 we model the operation OR (union).

And how then can we solve problems 1 and 2 in this case?

Answer 1 is max (0.1, 0.9) = 0.9 and answer 2 is $0.6 + 0.9 - 0.6 \times 0.9 = 0.96$

Now you see different problems may demand different solutions.

2.3 Extension principle and fuzzy algebra

2.3.1 Extension principle

OK. Now we understand how to perform logical operations. How do we make non-logical operations with fuzzy sets?

In any system, including control ones, different variables are interconnected with each other with algebraic formulas and equations. These relationships may reflect physical laws or a system structure. For example, we know that the voltage across a resistor is connected to the current with the formula: $I = V/R$ and to the power with the formula $W = V^2/R$. We want to know how our information about the voltage, expressed as a fuzzy set, can be used to get a model for the power and the current.

The question is: how to get a fuzzy model for a variable if we know the fuzzy model for another variable and the functional relationship between them. The answer is the extension principle, one of the fundamentals of fuzzy sets theory.

What do we need this extension principle for?

It gives us the rule of how to calculate an output of a fuzzy system. If we know the structure of the system, containing algebraic and logical blocks, and the system inputs are fuzzy, on the basis of this principle we can determine the outputs of the system.

> Definition 2.11 Extension principle 1.
>
> If A is a fuzzy set in the universe U and f is a mapping from U to the universe Y, y = f(u) then the extension principle allows us to define a fuzzy set B in Y as
>
> $$B = f(A) = \{(y, \mu_B(y))|_{y = f(u),\, u \in U}\}$$
>
> where
>
> $$\mu_B(y) = \begin{cases} \sup\limits_{u \in f^{-1}(y)} \mu_A(u) & \text{if } f^{-1}(y) \neq 0, \\ 0 & \text{if } f^{-1}(y) = 0 \end{cases}$$

The reason for such a definition is quite obvious. We should expect the same fuzziness incorporated in the second universe as there was in the first universe.

I understand that. But why do we apply the supremum operation?

Because the function f may map different elements of the universe U into one element of the universe Y. As a result this element may inherit a few membership degrees, hence, we need to determine one and we choose the highest one.

Example 2.5 *Let* $A = \{-2/0.3, -1/0.4, 0/0.8, 1/1, 2/0.7\}$ *and* $y = f(u) = u^2$.
Then $B = \{4/0.3, 1/0.4, 0/0.8, 1/1, 4/0.7\} = \{0/0.8, 1/1, 4/0.7\}$.

It is rather simple when we deal with a single-input-single-output (SISO) system. What will happen if we take a system with a few inputs, for example, a structure as Fig. 2.13?

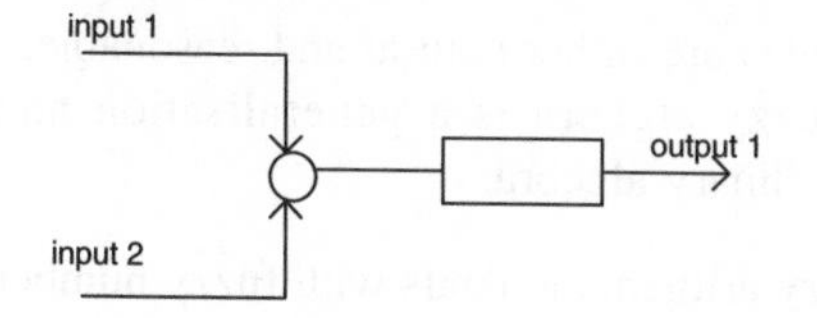

Fig. 2.13 *Two-input-one-output system structure*

In this case we need a more general form of the extension principle.

Definition 2.12 Extension principle 2.

Let X be a Cartesian product of the universes $X = U_1 * U_2 * \ldots * U_r$, and $A_1, A_2, \ldots, A_r$ are fuzzy sets in these universes. f is a mapping from X to the universe Y where $y = f(u_1, u_2, \ldots, u_r)$. Then the extension principle allows us to define the fuzzy set B in Y as

$$B = \{(y, \mu_B(y))|_{y = f(u1, u2, \ldots, ur), (u1, u2, \ldots, ur) \in X}\}$$

where

$$\mu_B(y) = \begin{cases} \sup \min \{\mu_{A1}(u_1), \ldots, \mu_{Ar}(u_r)\} \text{ if } f^{-1}(y) \neq 0, \\ (u_1, u_2, \ldots, u_r) \in f^{-1}(y) \\ 0 \qquad \text{if } f^{-1}(y) = 0 \end{cases}$$

Could you explain in greater detail, how this principle works, please?

Let us first consider just a case of two variables: the sum $y = u_1 + u_2$. Then we have

$$\mu_B(y) = \sup_{y = u_1 + u_2} \min \{\mu_{A1}(u_1), \mu_{A2}(u_2)\}$$

One can avoid considering an inverse operation. It means that for any element y of the universe Y we can take any couple of the elements u_1 of U_1 and u_2 of U_2, the sum of which equals to y and determine the membership degree as the minimum of the corresponding degrees. You see that one may have a huge or even infinite number of combinations of two numbers with the same sum and one needs to choose the supremum of all minimums.

It is rather difficult to apply. Can we use something simpler?

In a case of algebraic operations and some classes of fuzzy sets, we can apply fuzzy algebra and fuzzy numbers. Fuzzy algebra is not difficult to learn because:

- rules are rather natural and reasonable,
- fuzzy algebra is a generalisation and an extension of an ordinary algebra.

Fuzzy arithemetic deals with fuzzy numbers.

2.3.2 Fuzzy numbers

Definition 2.13 Fuzzy number.

If the support of the membership function is a part of the real axis, the fuzzy set represents a fuzzy number if the membership function is normal and convex.

In control and other physical systems we deal with measured (compared against the measurement scale) inputs and outputs. One can use fuzzy numbers to model real system variables. Classical arithmetic and algebra deal with crisp (ordinary) numbers.

Can you write the membership function for the crisp number, please?
Yes. For the number A it will be

$$\mu(x) = \begin{cases} 1, & x = A \\ 0, & x \neq A \end{cases}$$

Graphically this membership function can be presented as in Fig. 2.14.

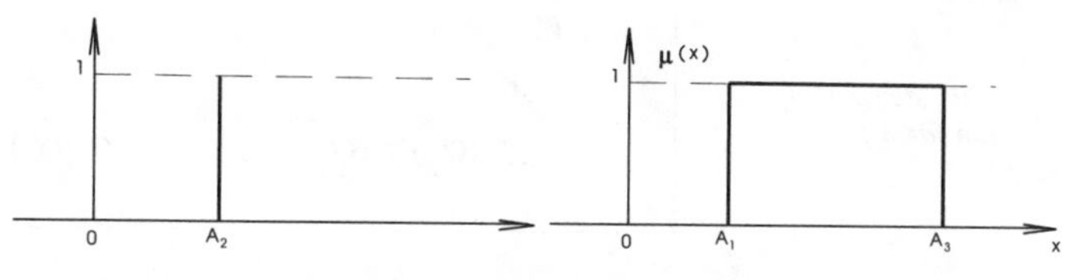

Fig. 2.14. *Membership function for a crisp number*

Fig. 2.15. *Membership function for an interval*

And what about the interval? I mean when we know exactly that the value cannot be smaller than A_1 and greater than A_3.

If we have an interval, the membership function can be written as

$$\mu(x) = \begin{cases} 0, & x < A_1 \\ 1, & A_1 \leq x \leq A_3 \\ 0, & x > A_3 \end{cases}$$

or graphically as in Fig. 2.15

Now let us consider a general fuzzy number (Fig. 2.16)

Before we start I want to ask about normal and convex fuzzy sets. We applied these words in the definition of a fuzzy number. What do they mean?

Definition 2.14 Normal fuzzy set.
The fuzzy set is normal if max $\mu(x) = 1$.

Definition 2.15 Convex fuzzy set
Convex means that any α–cut which is parallel to the horizontal axis

$$A\alpha = [a_1(\alpha), a_3(\alpha)]$$

yields the property of nesting, that is if

$\alpha' < \alpha \;=>\; a_1(\alpha') \leq a_1(\alpha)$ and $a_3(\alpha') \geq a_3(\mu\alpha)$.

Alternatively (see Fig. 2.16), if we represent the α-cut by the interval Aα as
$A\alpha = [a1(\alpha), \; a3(\alpha)]$ and $A\alpha' = [a1(\alpha'), a3(\alpha')]$
then the condition of convexity implies that if $\alpha' < \alpha \;=>$ $A\alpha \subset A\alpha'$.

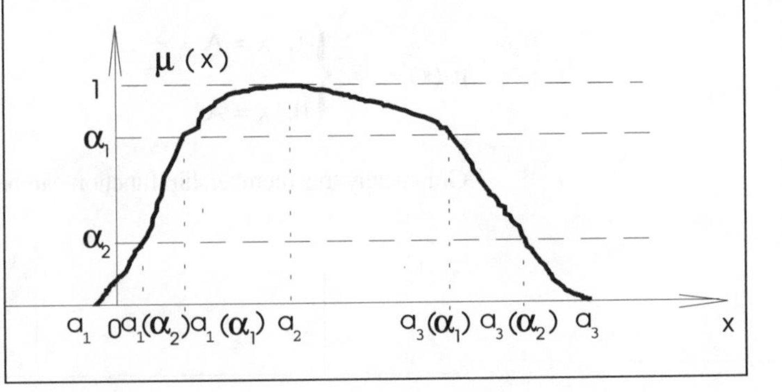

Fig. 2.16 *Membership function for a fuzzy number*

You see that the fuzzy set can be considered as a series of α-cut-sets which are represented by the intervals determined on the real axis. For the convex normal fuzzy sets these intervals are nested.

And what use is this?

It means that we can derive operations on fuzzy numbers based on operations with intervals. These intervals are called the intervals of confidence.

2.3.3 Arithmetic operations with intervals of confidence

Let me propose a problem: how would you find the resulting interval for some arithmetic operations with the intervals? Any idea?

Based on the extension principle we should find the interval which covers all possible results for all possible combinations of the operands.

That's right! Look at the following:

Addition: $[a_1, a_3] + [b_1, b_3] = [a_1 + b_1, a_3 + b_3]$

Subtraction: $[a_1, a_3] - [b_1, b_3] = [a_1 - b_3, a_3 - b_1]$

What is $[-2,4] + [3,5]$?

The left boundary will be the sum of the left boundaries of the operands and the right boundary will equal to the sum of the right boundaries. So $[-2,4] + [3,5] = [1,9]$

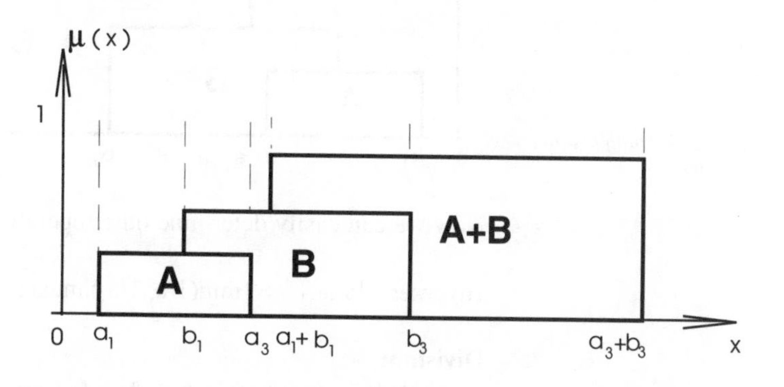

Fig. 2.17 *Addition of intervals*

Good! And in the case of subtraction?

The left boundary of the result will be a difference between the left boundary of the first operand and the right boundary of the second one and the right boundary will equal the difference between the right boundary of the first operand and the left boundary of the second one.

Very good! Now we are ready to consider multiplication.

Multiplication: $[a_1, a_3] \times [b_1, b_3] = [a_1 \times b_1, a_3 \times b_3]$? (2.3)

For example $[3,5] \times [4,6] = [12,30]$.

However, if you consider $[3,5] \times [-6, -4] = [-18, -20]$ you are not able to say that equation (2.3) is valid. A left boundary cannot be larger than a right one.

In this case, the result will be $[-20, -18]$, *we need just to swap the boundaries.*

Your resulting interval is very narrow, much narrower than in the case of all positive operands. It is confusing, isn't it?

Yes, it is.

The equation (2.3) is valid only if $a_1, b_1 \geq 0$, i.e. both operands are non-negative.

Generally, we should apply the formula

$[a_1, a_3] \times [b_1, b_3] = [\min(a_1 \times b_1, a_1 \times b_3, a_3 \times b_1, a_3 \times b_3), \max(a_1 \times b_1, a_1 \times b_3, a_3 \times b_1, a_3 \times b_3)]$

which is always right.

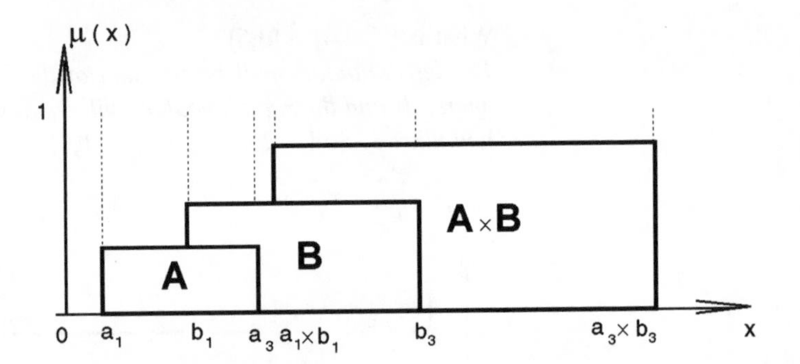

Fig. 2.18 *Multiplication of intervals*

Now we can easily determine other operations with the intervals.

Inverse: $[a_1,a_3]^{-1} = [\min(1/a_1,1/a_3),\max(1/a_1,1/a_3)]$.

Division:
$[a_1,a_3]/[b_1,b_3] = [\min(a_1/b_1,a_1/b_3,a_3/b_1,a_3/b_3), \max(a_1/b_1,a_1/b_3,a_3/b_1,a_3/b_3)]$.

If a fuzzy set is defined over the set of positive real numbers only (that is we consider just positive operands), the last formulas can be simplified:

$[a_1,a_3]^{-1} = [1/a_3,1/a_1]$

$[a_1,a_3]/[b_1,b_3] = [a_1/b_3,a_3/b_1]$

If we denote the max operation as $\cup$ and a min operation as $\cap$, we can rewrite the formulas given above as

$[a_1,a_3] \times [b_1,b_3] = [a_1 \times b_1 \cap a_1 \times b_3 \cap a_3 \times b_1 \cap a_3 \times b_3,\ a_1 \times b_1 \cup a_1 \times b_1 \cup a_1 \times b_3 \cup a_3 \times b_1 \cup a_3 \times b_3]$

$[a_1,a_3]^{-1} = [1/a_1 \cap 1/a_3,\ 1/a_1 \cup 1/a_3]$

$[a_1,a_3] / [b_1,b_3] = [a_1/b_1 \cap a_1/b_3 \cap a_3/b_1 \cap a_3/b_3,\ a_1/b_1 \cup a_1/b_3 \cup a_3/b_1 \cup a_3/b_3]$

Why have we replaced one notation with another one?

We are trying to prepare for an introduction of operations with fuzzy numbers.

2.3.4 Arithmetic operations with fuzzy numbers

You remember that a fuzzy set A can be considered as a union of α-cut-sets:

$$A = \bigcup_{\alpha \in [0,1]} A\alpha$$

The results obtained above for the intervals can be expanded to fuzzy numbers A and B by expressing the fuzzy numbers at each level of α as an interval of confidence.

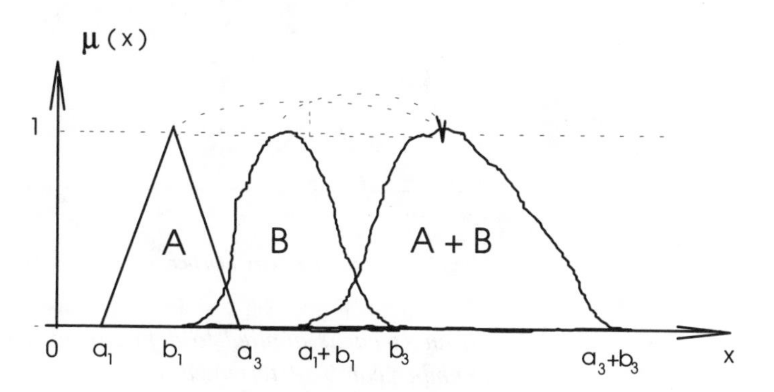

Fig. 2.19 *Addition of fuzzy numbers*

The following formulas can be derived:

Addition: $\mu_{A+B}(z) = \bigcup_{z=x+y} (\mu_A(x) \cap \mu_B(y))$

Subtraction: $\mu_{A-B}(z) = \bigcup_{z=x-y} (\mu_A(x) \cap \mu_B(y))$

Multiplication: $\mu_{A\times B}(z) = \bigcup_{z=x\times y} (\mu_A(x) \cap \mu_B(y))$

Division: $\mu_{A/B}(z) = \bigcup_{z=x/y} (\mu_A(x) \cap \mu_B(y))$

Are these formulas valid for any fuzzy numbers?

Yes, they are, including a special case of fuzzy numbers – crisp or ordinary numbers. One can replace any fuzzy number with an ordinary one, giving ordinary arithmetic.

Actually calculation of these formulas requires a lot of work. How can I decrease the time necessary for computation?

We can consider some special cases of fuzzy numbers which are the most commonly used in applications. In these cases, the situation can be significantly simplified.

Triangular fuzzy numbers

This class of fuzzy numbers has been used extensively in different applications, first of all because of its simplicity. The membership function is

$$\mu(x) = \begin{cases} 0, & x < a_1 \\ (x - a_1)/(a_2 - a_1), & a_1 \leq x \leq a_2 \\ (a_3 - x)/(a_3 - a_2), & a_2 \leq x \leq a_3 \\ 0, & x > a_3 \end{cases}$$

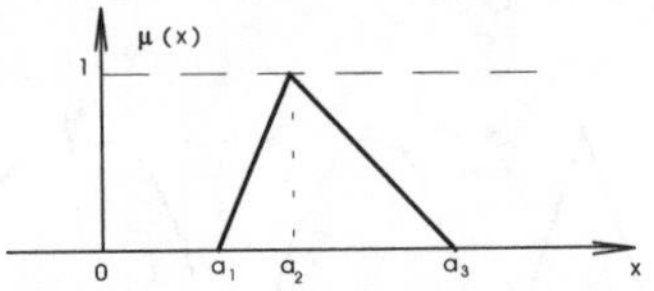

Fig. 2.20 *Triangular fuzzy number*

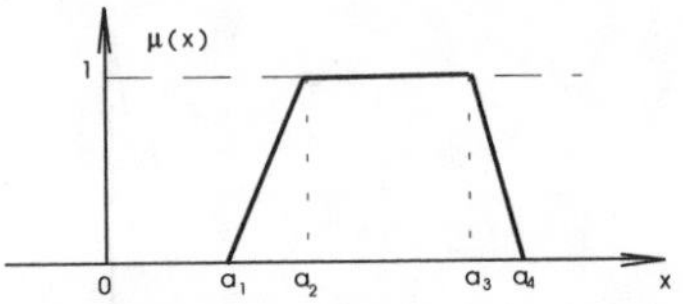

Fig. 2.21 *Trapezoidal fuzzy number*

If an α–cut is applied to define a fuzzy number then the fuzzy number can be determined as

$\forall \alpha \in [0,1]$: $A\alpha = [a_{1\alpha}, a_{3\alpha}] = [(a_2 - a_1) \times \alpha + a_1, -(a_3 - a_2) \times \alpha + a_3]$

So a triangular fuzzy number (TFN) can be given by three ordinary numbers (a_1, a_2, a_3).

Arithmetic operations on TFN have the following properties:

- Addition or subtraction operations on two TFN result in another TFN;
- Multiplication, inversion and division operations on TFN do not necessarily give a TFN as a result.

Addition: $A + B = (a_1,a_2,a_3) + (b_1,b_2,b_3) = (a_1+b_1,a_2+b_2,a_3+b_3)$
Subtraction: $A - B = (a_1,a_2,a_3) - (b_1,b_2,b_3) = (a_1-b_3,a_2-b_2,a_3-b_1)$

Trapezoidal fuzzy numbers (TrFN)

The membership function is:

$$\mu(x) = \begin{cases} 0, & x < a_1 \\ (x - a_1)/(a_2 - a_1), & a_1 \leq x \leq a_2 \\ 1, & a_2 \leq x \leq a_3 \\ (a_3 - x)/(a_3 - a_2), & a_3 \leq x \leq a_4 \\ 0, & x > a_4 \end{cases}$$

A TrFN can be described completely by a quadruplet $A = (a_1,a_2,a_3,a_4)$. It can also be characterised by the interval of confidence at level α. Thus,

$$\forall \alpha \in [0,1]: \quad A\alpha = [\, a_{1\alpha}, a_{4\alpha} \,] = [(a_2-a_1) \times \alpha + a_1, \; -(a_4-a_3) \times \alpha + a_4]$$

As one can see, a TFN is a particular case of a TrFN with $a_2 = a_3$.

Properties:

- addition or subtraction operations on two TrFN give another TrFN;
- multiplication, inversion and division operations on TrFN do not necessarily give a TrFN.

Addition: $A + B = (a_1,a_2,a_3,a_4) + (b_1,b_2,b_3,b_4) = (a_1+b_1,\; a_2+b_2,\; a_3+b_3,\; a_4+b_4)$
Subtraction: $A - B = (a_1,a_2,a_3,a_4) - (b_1,b_2,b_3,b_4) = (a_1-b_4,\; a_2-b_3,\; a_3-b_2,\; a_4-b_1)$

Gaussian Fuzzy Numbers (GFN)

$$\mu(x) = \begin{cases} \exp\left(-((x-m)/\sigma_1)^2\right), & x \le m \\ \exp\left(-((x-m)/\sigma_2)^2\right), & x > m \end{cases}$$

GFN can be described by a triplet (m, σ_1, σ_2) where m is the crisp magnitude of the GFN and σ_1, σ_2 are fuzziness parameters (a membership function is given in Fig. 2.5c).

Arithmetic operations on GFN have the following properties:

- Addition or subtraction operations on two GFN give a GFN.
- Addition: $A + B = (m_a, \sigma_{1a}, \sigma_{2a}) + (m_b, \sigma_{1b}, \sigma_{2b}) = (m_a+m_b, \sigma_{1a}+\sigma_{1b}, \sigma_{2a}+\sigma_{2b})$
- Subtraction: $A - B = (m_a, \sigma_{1a}, \sigma_{2a}) - (m_b, \sigma_{1b}, \sigma_{2b}) = (m_a - m_b, \sigma_{1a}+\sigma_{1b}, \sigma_{2a}+\sigma_{2b})$
- Multiplication, inversion and division operations do not generally produce a GFN.

However, if the fuzziness parameters σ_1 and σ_2 are much less than the magnitudes of the operands, the result can be described as a GFN. In this case one can consider $\sigma_1 = \sigma_2$ and will have for the multiplication:

$$A \times B = (m_a, \sigma_{1a}, \sigma_{2a}) \times (m_b, \sigma_{1b}, \sigma_{2b}) \approx (\, m_a \times m_b, \sigma_a \times \text{abs}(m_a) + \sigma_b \times \text{abs}(m_b), \sigma_a \times \text{abs}(m_a) + \sigma_b \times \text{abs}(m_b))$$

2.4 Linguistic variables and hedges

I have heard that fuzzy logic operates with words and terms of natural language and you have been presenting just mathematical formulas.

Fuzzy sets theory operates just with mathematical models as any other mathematical theory does. It replaces one sort of mathematical model with another one. However, you are right, fuzzy sets theory allows us to model terms of natural language with the help of linguistics variables. It has opened a wide area of use of this theory in control and other applications.

How does it work?

A linguistic variable is one with a value that is a natural language expression referring to some quantity of interest. These natural language expressions are then in turn names for fuzzy sets composed of the possible numerical values that the quantity of interest can assume.

So is the value of the linguistic variable a word?

Yes, a word or a word sentence. This is the main difference between a linguistic variable and a numerical one.

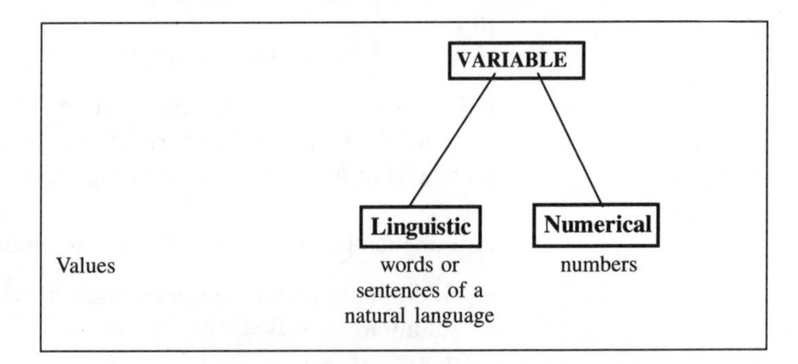

Generally, a linguistic variable is a composite term $u = u_1, u_2, ...,$ u_n which is a concatenation of atomic terms $u_1, u_2, ..., u_n$, and according to [Cox94] these atomic terms can be divided into four categories:

- primary terms, which are the labels of specified fuzzy subsets of the universe of discourse (e.g. small and big);
- the connectives AND, OR and the negation NOT;
- hedges such as VERY, MOST, RATHER, SLIGHTLY, MORE OR LESS, etc.;
- markers such as parenthesis.

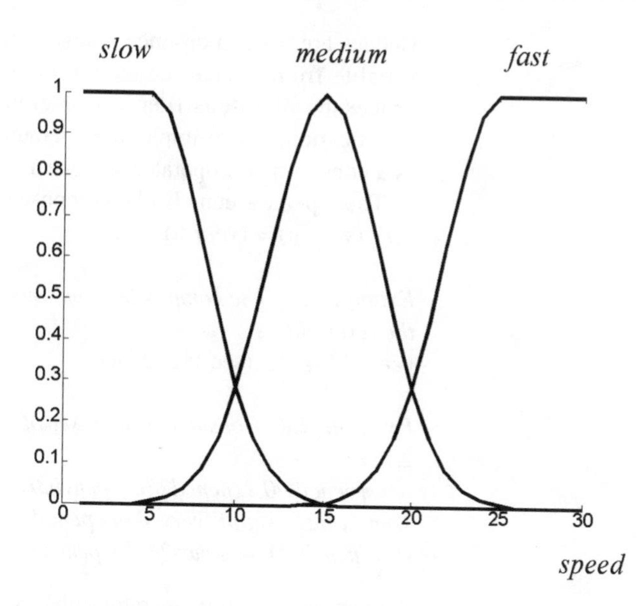

Fig. 2.22 *Membership functions for linguistic values*

A primary term is actually just a name or a label of a fuzzy set. It usually describes the word which is used by experts to express their opinion about the value of one of the object characteristics, e.g old, large, fast, etc.

The connectives realise the operations of intersections, union, complement considered earlier.

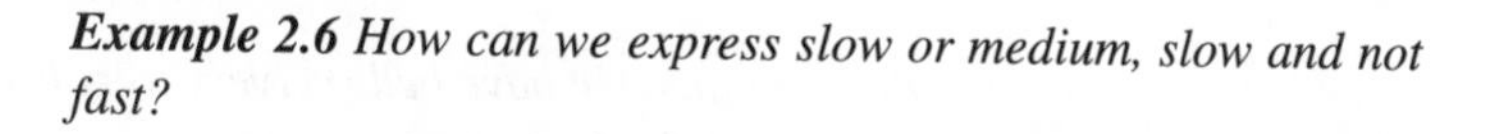

Example 2.6 *How can we express slow or medium, slow and not fast?*

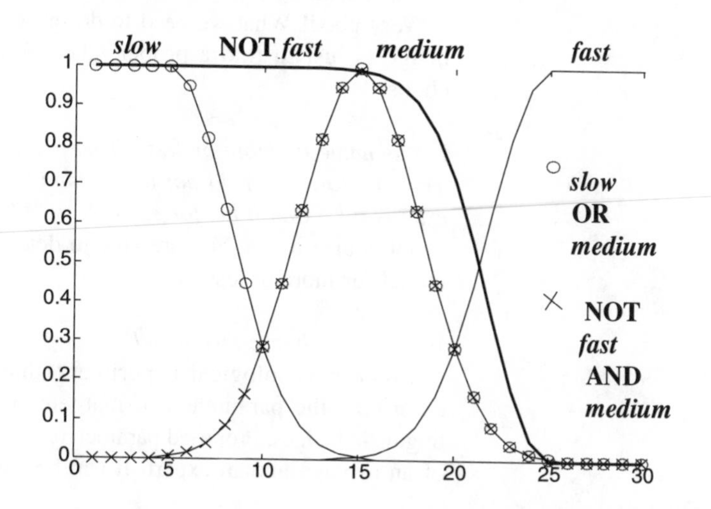

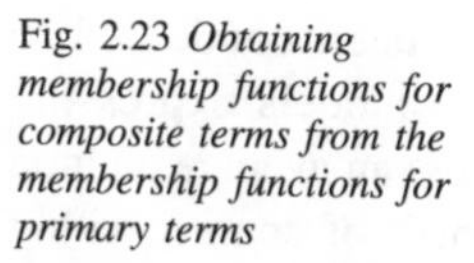

Fig. 2.23 *Obtaining membership functions for composite terms from the membership functions for primary terms*

Hedges are used to produce a larger set of values for a linguistic variable from a small collection of primary terms through the processes of intensification or concentration, dilation and fuzzification. For example, the operator '*very*' is usually defined as a concentration operator as *very* u = u^2

This operator can also be composed with itself, thus *very* (very u) = $(\text{very u})^2 = u^4$

Example 2.7 *The composite term 'very old' can be obtained from the term old as*
very old = old^2 (see Fig. 2.24)

Example 2.8 *Consider a fuzzy set of 'short pencils' A (Example 2.1) as*
A = {pencil1/0.2, pencil2/0.5, pencil3/1.0, pencil4/1.0, pencil5/0.9}.
Then a fuzzy set of 'very short pencils' can be determined as
B = {pencil1/0.04, pencil2/0.25, pencil3/1.0, pencil4/1.0, pencil5/0.81}.

We can see that this operator enhances an action of the term. It significantly changes the membership function shape. Try to find the inverse hedge, which 'fuzzifies' the term and expands the width of the corresponding membership function.

Obviously this hedge (let us denote it now as hedg1) can be written as hedg1 $u = u^{1/2}$

Why do you think so?

Because then we have hedg1 (very u) = hedg1 $(u^2) = (u^2)^{1/2} = u$ *producing the inverse action.*

Very good! What we need to do in order to form a linguistic hedge is just to find a proper label. What would you like to choose?

Let us name it 'more or less'. Then we have more or less (very u) = u. However, I do not understand why more or less means exactly $u^{1/2}$. *Can it be, for example,* $u^{0.4}$*?*

Of course it can. You are able to determine the mathematical model for more or less as you wish.

Why do we choose this model?

Some psychological experiments have been conducted to determine the parameters of mathematical models expressing linguistic hedges. Proposed parameters express an average opinion of an average human expert. It can be varied, of course, within the borders limited by just a common sense. You need to remember just that a hedge changes a membership degree for any part in the domain or we can say in the scientific language that it transforms a fuzzy surface.

A scientific approach features not just a specific language but also an attempt to classify objects. Can we classify all possible hedges?

I do not think we can. As hedges could be constructed from words and expressions of the natural language and a vocabulary of the natural language is immense we are able to describe a very limited subset of all possible hedges. And as we discussed earlier every user could construct the mathematical description for any hedge which is applied in any model. [Cox94] classified some widely used hedges (see Table 2.3).

Table 2.3 Some widely used hedges

HEDGE	MEANING
about, around, near, roughly	approximate a scalar
above, more than	restrict a fuzzy region
almost, definitely, positively	contrast intensification
below, less than	restrict a fuzzy region
vicinity of	approximate broadly
generally, usually	contrast diffusion
neighbouring, close to	approximate narrowly
not	negation or complement
quite, rather, somewhat	dilute a fuzzy region
very, extremely	intensify a fuzzy region

There are different types of hedges in that table. Which one is the most important?

It is not easy to answer that question. Hedges represent the words of natural language. Which words are more important than others? Moreover, this table does not cover all hedges and all words. We have included just a small subset of hedges and words, maybe the most widely used. For example, approximation hedges play an important role because they fuzzify a scalar.

What do you mean? How can we fuzzify a scalar number?

I mean that we can obtain the fuzzy domain for the variable. For example 'height is about 3 m' will produce the following fuzzy set (Fig. 2.25). The role of the approximation hedges is to convert scalar values into fuzzy values and their place is at the beginning of the fuzzification process.

Why did you choose this membership function? And generally how does an approximation hedge know how wide to make the support of the fuzzy set?

An important question, especially because the approximation hedges usually begin the process and determine the consequent results. To choose the width we should consider these results: how does the domain of the generated fuzzy region match the domain of the consequent fuzzy space. We have to try to determine this domain from our knowledge of the problem or the problem context. We should consider not just immediate results but also consequent fuzzy sets which are determined on the base of the initial ones.

Why do you constantly repeat the phrase 'consequent results'?

Let me explain. If we want to determine the domain for the fuzzy variable 'about 5' and we know that it will be used to determine the pressure at the testing point 1 and we know the rule that 'if the pressure at the testing point 1 is about 5 then the temperature at the testing point 23 is high' and we know the domain for this temperature, then we can consider how different domains for the pressure match this known domain for the temperature.

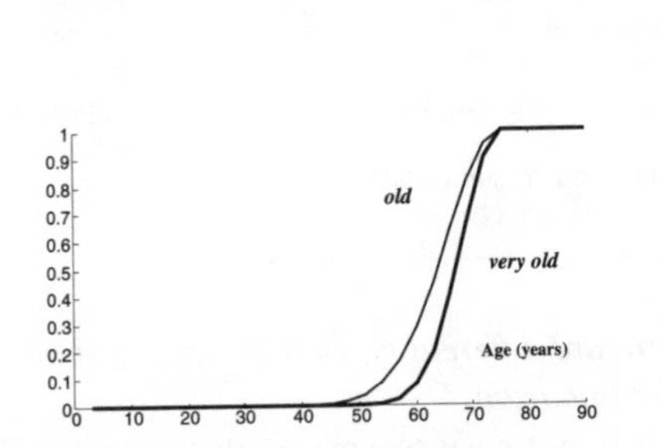

Fig. 2.24 Obtaining the membership function for the term 'very old' from the membership function for the term 'old'

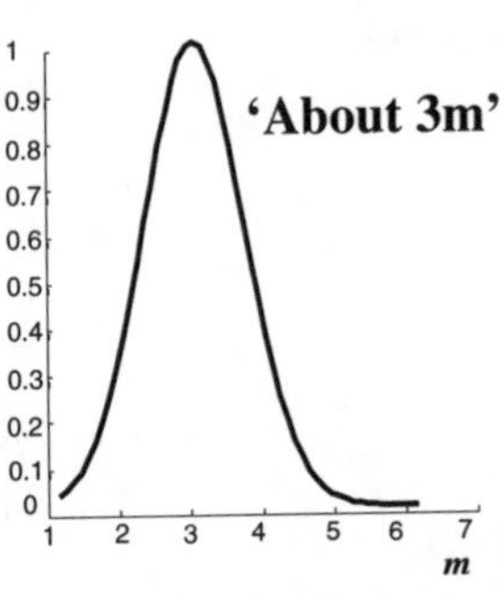

Fig. 2.25 Membership function 'about 3 m'

So we should try the membership function, consider the results and try again?

Yes, this is a usual design process, consisting of tries and errors.

If we already have the fuzzy set, for example 'the speed should be around average' (Fig. 2.26) can we apply the approximation hedges here?

Sure we can. We shall fuzzify the previous domain a little bit.

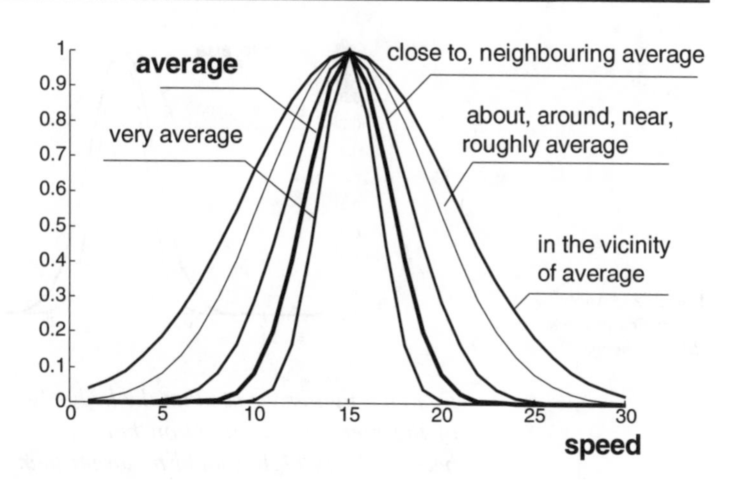

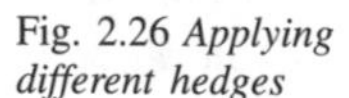
Fig. 2.26 *Applying different hedges*

What do other hedges do?

The hedges 'about', 'around', 'near', 'roughly' perform the 'average' fuzzification of the domain. The hedges like 'close to', 'neighbouring' perform a narrow approximation or fuzzification. However, the hedge 'in the vicinity of' expands the initial domain significantly (Fig. 2.26). Be careful with this hedge! Its application could result in a fuzzy set with the domain exceeding the general domain of the working fuzzy set and could lead to an unforeseen degree of ambiguity.

What else can the hedge perform except fuzzifying the domain of the fuzzy set?

A lot of different things. Restrict a fuzzy domain, for example, like hedges 'above', 'more than', 'below', 'less than' .

Can we apply these hedges to a scalar? How do we construct a membership function 'above 5'?

[Cox94] proposes a crisp set where we have the step function with the border exactly at the level of 5 . However, if we initially apply the approximation hedge and the restriction hedge 'above around 5', we have the usual membership function. An important class of the hedges are those which intensify a fuzzy region ('very', 'extremely') or dilute it ('quite', 'rather', 'somewhat'). The intensification hedges reduce a membership degree for each value of the domain, while the dilution hedges increase it (Fig. 2.27).

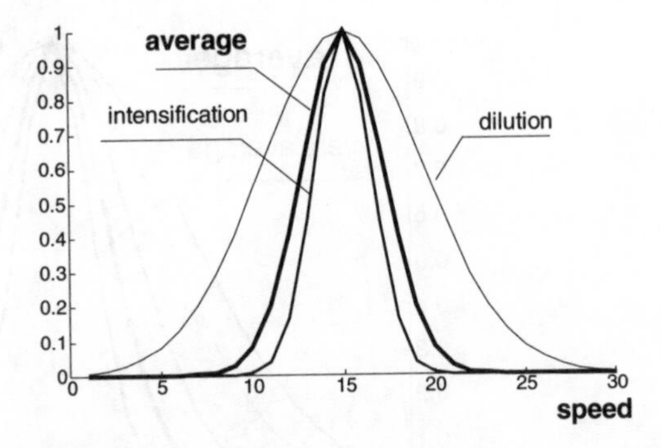

Fig. 2. 27 *Applying intensification and dilution hedges*

From the figures we see that the hedge 'very' decreases the value of the membership function but I think it should increase it, because 'very high' should be larger than 'high'.

That's right! The hedge decreases a value of the membership function for any value in the domain of, for example, a man's height. For example, 180 cm can be considered as 'high' with the degree of 0.75 and 'very high' with the degree 0.57. On the other hand, to be considered as 'very high' with the same degree of 0.75, a man should have a height of 190 cm. So the hedge application definitely has increased the value of the height, which has the same membership degree (Fig. 2.28).

Can we use more than one hedge?

Yes. The most important thing is that we can combine different hedges, building rather complex linguistic expressions, for example 'almost very fast but *generally* below 100 km/h' or 'close to 100 m but not very high'.

Sounds good. So we can use almost all expressions of a natural language to describe something.

That's right.

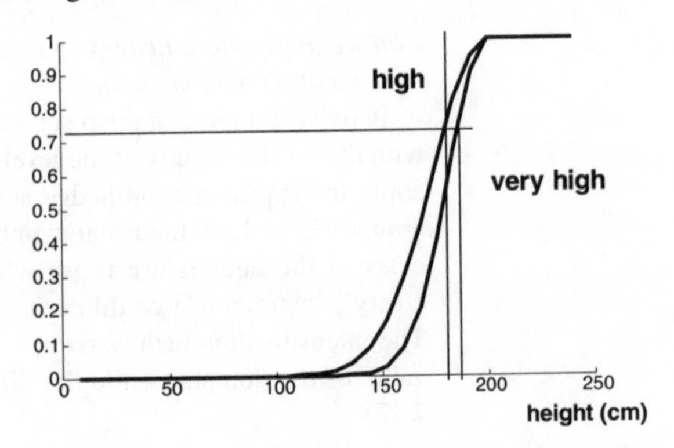

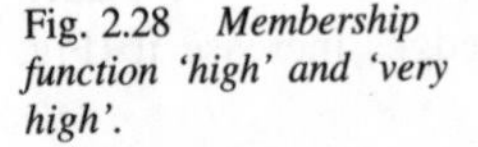
Fig. 2.28 *Membership function 'high' and 'very high'.*

But how can a computer understand and process these expressions?

Hedges in complex expressions are processed in a manner analogous to English adjectives. Thus the expression 'not more than above zero' is interpreted as not (more than (about(zero))). Firstly, a fuzzy region 'about zero' is created. Note, that it is constructed by applying the hedge 'about' to the crisp number of zero. The main thing you should remember is that the linguistic variables and hedges allow us to construct mathematical models for expressions of natural language. These models can be used to write and process rules and other objects.

2.5 Fuzzy relations

Now, when we know how to determine a fuzzy set and what it is, we are able to consider a relationship between fuzzy sets or the fuzzy relations. The fuzzy relation regards at least two fuzzy sets.

What do we need this fuzzy relation for?

We need it to understand and formalise application problems. Many application problem descriptions include fuzzy relations. For example, to describe a plant or a control system one determines, how an output(s) depends on inputs, or the relationship between outputs and inputs. If one constructs a database and an information system, one determines the relations between different attributes. To model a fuzzy system one uses rules like

if speed is *slow* ***then*** pressure should be *high*

If the speed is denoted as variable A and pressure as variable B then one will have in a general case the rule: ***if*** A ***then*** B

This rule gives the dependence of output B on input A. How do we determine the output if we know the input?

We must know the relation between these two variables.

Definition 2.16 A fuzzy relation of the form $A \Rightarrow B$, R, can be defined as a relation of two fuzzy sets $A \in U$ and $B \in V$ as a fuzzy subset on the Cartesian product U *V. R is characterised by the bivariate membership function $\mu_R(u, v)$ as

$$R = A * B = \int_{U*V} \mu_R(u, v) / (u,v)$$

or for finite sets:

$$R = A * B = \Sigma_{u,v} \mu_R(u, v) / (u,v) \qquad (2.4)$$

This is very difficult for non-mathematical people to understand. What should we integrate?

Actually we do not integrate anything in a sense of a mathematical operation.

What does the $\int$ sign mean then?

It shows that all possible combinations of all the elements of both universes of discourse should be considered. It is easier to explain when both universes are finite sets. Suppose we have two finite fuzzy sets: A which is given by its membership function μ_A = (1/1, 0.8/2, 0.6/3, 0.5/4, 0.4/5, 0.2/6), and B with the membership function μ_B = (0.5/5, 1/6, 0.3/7). Now we should consider all the combinations of the elements of the universes where the membership function has non-zero values. How many combinations do we have here?

The first fuzzy set has 6 non-zero elements and the second one has 3. Then the fuzzy relation has 6 × 3 = 18 elements. Here are these combinations: (1,5), (1,6), (1,7), (2,5), (2,6), (2,7), (3,5), (3,6), (3,7), (4,5), (4,6), (4,7), (5,5), (5,6), (5,7), (6,5), (6,6), (6,7).

That's right! And now we need to determine a membership degree for every pair.

How?

Just use the definition! You can use either the minimum or product. If we use the minimum we have

μ_R (u,v) = {0.5/ (1,5), 1/(1,6), 0.3/(1,7), 0.5/(2,5), 0.8/(2,6), 0.3/(2,7), 0.5/(3,5), 0.6/(3,6), 0.3/(3,7), 0.5/(4,5), 0.5/(4,6), 0.3/(4,7), 0.4/(5,5), 0.4/(5,6), 0.3/(5,7), 0.2/(6,5), 0.2/(6,6), 0.2/(6,7)}.

And what will happen if we use the product?

μ_R (u,v) = {0.5/(1,5), 1/(1,6), 0.3/(1,7), 0.4/(2,5), 0.8/(2,6), 0.24/(2,7), 0.3/(3,5), 0.6/(3,6), 0.18/(3,7), 0.25/(4,5), 0.5/(4,6), 0.15/(4,7), 0.2/(5,5), 0.4/(5,6), 0.12/(5,7), 0.1/(6,5), 0.2/(6,6), 0.06/(6,7)}

We have different answers!

Again different problems can demand different solutions! And in the case of continuous membership functions, we use the sign $\int$ to show 'integration' of all possible combinations.

Hang on! The fuzzy relation is the same fuzzy set but determined on the universe which is a combination of two other universes.

Absolutely! To denote this new universe we use the words 'Cartesian product' to demonstrate that we formed a new support set.

Why did we use just the minimum and product operations? Can we use other operations from Table 2.2?

Sure we can! More accurately we should have determined the equation (2.4) as a $\hat{t}$ norm operation.

$$\mu_R(u,v) = \int_{u,v} \mu_A(u)\ \hat{t}\ \mu_B(v) / (u,v)$$

or

$$\mu_R(u,v) = \Sigma_{u,v} \mu_A(u)\ \hat{t}\ \mu_B(v) / (u,v).$$

To choose which one to use, let us compare their actions. You can consider different applications described in the literature (see, for example, [Har93]). Even for simple illustrative examples, it can be observed that an algebraic product operator leads to a smoother fuzzy relation surface.

Now let us try to make some calculations of fuzzy relations.

Definition 2.17 N-ary fuzzy relation
Similarly to (2.4), we can determine the n-ary relation between n fuzzy sets.

$$R_{U_1 * U_2 * \ldots * U_n} = \int_{U_1 * \ldots * U_n} \mu_R(u_1, \ldots, u_n) / (u_1, \ldots, u_n) \in U_1 * \ldots * U_n$$

or in the case of discrete sets

$$R_{U_1 * U_2 * \ldots * U_n} = \Sigma_{U_1 * \ldots * U_n} \mu_R(u_1, \ldots, u_n) / (u_1, \ldots, u_n) \in U_1 * \ldots * U_n$$

where $u_1 \in U_1$, $u_2 \in U_2$, ..., $u_n \in U_n$.

You see that this relation is determined on the n-dimensional space $U_1 * \ldots * U_n$.

This is very difficult to understand! What does n-ary fuzzy relationship mean? How can I imagine it?

To imagine it you should have a multidimensional space vision.

How do you calculate its result?

Actually you need to calculate the membership function of this relation as a Cartesian product. To do it you may use any $\hat{t}$ norm. Let us determine the Cartesian product of the fuzzy sets.

Definition 2.18 Cartesian product
If $A_1, \dots, A_n$ are fuzzy sets in the universes of discourse $U_1, \dots, U_n$ then the Cartesian product of $A_1, \dots, A_n$ is a fuzzy set in the product space $U_1 * \dots * U_n$ with the membership function

$$\mu_{A_1 * \dots * A_n}(u_1, \dots, u_n) = \mu_{A_1}(u_1) \hat{t} \mu_{A_2}(u_2) \hat{t} \dots \hat{t} \mu_{A_n}(u_n)$$

What do we need it for? I guess a simple two–dimensional fuzzy relation is more than enough.

With a two-dimensional relation you can derive the output of one simple rule: if A then B, e.g., if speed is *slow* then temperature is *low*. However, a fuzzy rule can consist of fuzzy sets from more than two disparate universes of discourse: if A then B then C, e.g., if speed is *slow* then temperature is *low* then pressure is *high*. In this case we should apply a n-ary relation. You have to remember that all the fuzzy sets here vary in nature and are determined in the different universal sets.

What will happen if we have a few rules? for example, if temperature is low then pressure is high, or, if temperature is medium then pressure is medium, or, if temperature is high then pressure is low.

A good question! Of course, in practical problems we usually have a few rules. To calculate the result we should combine all the rules. Mathematically every rule is described by a relation. All the rules will be described by the composition of these relations.

Definition 2.19 Multirelational composition
If R and S are fuzzy relations in the Cartesian spaces $U * V$ and $V * W$ respectively, the composition of R and S is a fuzzy relation given by the sup – $\hat{t}$ composition

$$\mu_{R \circ S} = \{[\sup_V (\mu_R(u,v) \hat{t} \mu_S(v,w))\}, u \in U, v \in V, w \in W\}$$

or by the inf-s composition

$$\mu_{R \circ S} = \{[\inf_V (\mu_R(u,v) s \mu_S(v,w))\}, u \in U, v \in V, w \in W\}.$$

I do not understand which formula to apply. We have got two of them.

If you look carefully you will see that we have more than two because $\hat{s}$ and $\hat{t}$ operations can be determined by different ways.

Your explanation has not made the situation easier. All the possible combinations are not used in practical applications, are they?

In practice the sup-min, sup-algebraic and if-max composition operators are utilised. And to make the situation easier you can show yourself, that sup-min and if-max operations produce the same result.

We can provide some properties of the sup-min composition:

$$(R \cup T) \circ S = R \circ S \cup T \circ S$$
$$\text{if } R_1 \in R_2 \text{ then } R_1 \circ S \in R_2 \circ S$$
$$(R \cap T) \circ S \in (R \circ S) \cap (T \circ S)$$
$$\text{for } R_1, R_2, R, T \in U * V, \quad \text{and } S \in V * W$$

In practice, the most important case of compositional inference is when we know the fuzzy input and the fuzzy relation between the input and the output and we need to obtain the output.

Definition 2.20 Compositional rule of inference
If R is a fuzzy relation in U * V and $\bar{\boldsymbol{A}}$ is a fuzzy set in U then the fuzzy set $\bar{\boldsymbol{B}}$ in V induced by $\bar{\boldsymbol{A}}$ is given by

$$B = \begin{cases} \bar{\boldsymbol{A}} \circ R \text{ for sup–t composition} \\ \bar{\boldsymbol{A}} \circ R \text{ for inf–s composition} \end{cases}$$

You see that because the fuzzy relation is actually the fuzzy set, we can determine the operations with fuzzy relations similarly as with fuzzy sets.

Definition 2.21
If R and S are fuzzy relations in U * V the intersection of R and S is

$$\mu_{R \cap S}(u,v) = \mu_R(u,v) \; \hat{t} \; \mu_S(u,v)$$

where t-norm can be calculated as a minimum operator, for example

$$\mu_{R \cap S}(u,v) = \min(\mu_R(u,v), \; \mu_S(u,v))$$

or a product $\mu_{R \cap S}(u,v) = \mu_R(u,v) \times \mu_S(u,v)$

Definition 2.22
If R and S are fuzzy relations in U * V the union of R and S is

$\mu_{R \cup S}(u,v) = \mu_R(u,v) \ \mathrm{s} \ \mu_S(u,v)$

where we can calculate $\hat{s}$ norm as a maximum operator, for example,

$\mu_{R \cup S}(u,v) = \max(\mu_R(u,v), \ \mu_S(u,v))$.

And what do we need these definitions for?

A fuzzy relation describes interactions between fuzzy variables. For example, if we have the statement 'Temperature 1 is about Temperature 2',we actually have two fuzzy variables: Temperature 1 and Temperature 2. By this statement we wish to describe the relation between them. Suppose Temperature 1 can have five values, 11, 12,13, 14, 15, and Temperature 2 can have three values, 10, 12, 14. Then the fuzzy relation can have the membership function:

μ_{R*S} (u,v) = (0.9/(10,11), 0.7/(10,12), 0.5/(10,13), 0.3/(10,14), 0.1/(10,15), 0.9/(12,11), 1/(12,12), 0.9/(12,13), 0.7/(12,14), 0.4/(12,15), 0.5/(14,11), 0.7/(14,12), 0.9/(14,13), 1/(14,14), 0.9/(14,15))

or if we rewrite it in a more convenient table (matrix) form:

		Temperature 1				
		11	12	13	14	15
Temperature 2	10	1.0	0.7	0.5	0.3	0.1
	12	1.0	1.0	1.0	0.7	0.4
	14	0.5	0.7	1.0	1.0	1.0

Consider now another statement: 'Temperature 1 seems to be larger than Temperature 2'. This fuzzy relation will have the membership function:

		Temperature 1				
		11	12	13	14	15
Temperature 2	10	0.9	1.0	1.0	1.0	1.0
	12	0.1	0.5	0.9	1.0	1.0
	14	0	0	0.1	0.5	0.9

Now we want to determine the membership function for the statement 'Temperature 1 is about Temperature 2 **and** Temperature

1 seems to be larger than Temperature 2.' This relation can be determined as an intersection of two relations and will have the membership function:

	Temperature 1					
		11	12	13	14	15
Temperature 2	10	1.0	0.7	0.5	0.3	0.1
	12	0.1	0.5	0.9	0.7	0.4
	14	0	0	0.1	0.5	0.9

Try to determine the membership function for the statement 'Temperature 1 is about Temperature 2 **or** Temperature 1 seems to be larger than Temperature 2.'

I understand this relation to be determined as a union of both and will have the membership function:

	Temperature 1					
		11	12	13	14	15
Temperature 2	10	0.9	1.0	1.0	1.0	1.0
	12	1.0	1.0	1.0	1.0	1.0
	14	0.5	0.7	1.0	1.0	1.0

The definitions and examples considered above allow us to obtain the fuzzy output of the fuzzy system block and to understand the basics of fuzzy reasoning methods.

3 THE STRUCTURE AND OPERATION OF A FUZZY CONTROLLER

3.1 The reasons to apply fuzzy controllers

During our study we have learned many techniques for the synthesis of automatic controllers. Why do we need to learn one more?

The proposed method differs significantly from conventional ones. Usually a control strategy and a controller itself is synthesised on the base of mathematical models of the object or process under control. The models of an object under control involve quantitative, numeric calculations and commonly are constructed in advance, before realisation.

What about adaptive or self-organising controllers?

In these controllers, the exact control strategy is not calculated in advance but is generated by optimisation algorithms based on the controller 'experience'. However, this approach also supposes the construction of 'numeric' mathematical models.

Isn't there another approach?

Yes, there is. A different method is to comprehend instructions and to generate strategies based on *a priori* verbal communication. Most control engineers would accept intuitively that mathematical modelling, which they perform in translating their concept of a control strategy into an automatic controller, is completely different from their own approach to a manual performance of the same task. On the other hand, linguistic description of control seems to be similar to its manual implementation.

I do not understand the reason for this. I guess if we apply a right method it should produce the best results (in some sense). I think people just apply the wrong numeric models.

I agree with you just partly. The problem is that often it is very difficult to derive the 'right' model. Many real objects and real life processes are too complicated to model mathematically. Instead, if we try to construct a good model, it may be too complex for a successful application of a control theory. In Fig. 3.1 you may see that there is a long way to go to solve a control problem. Unfortunately, one meets a lot of disturbances on the way. Usually even the best model is not complicated enough for a comprehensive description of a real process. On the other hand, it may be too complex to apply.

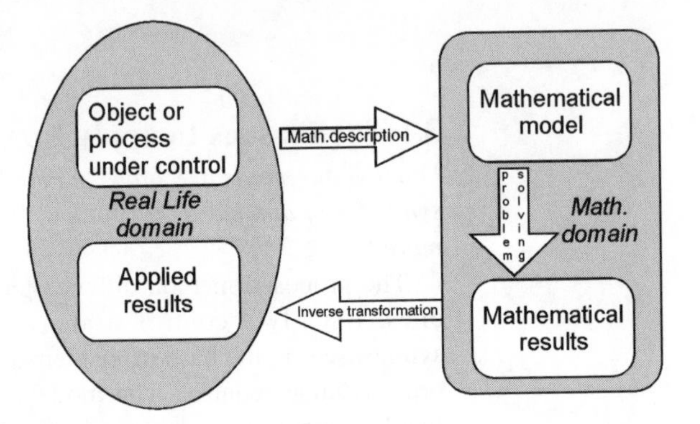

Fig. 3.1 *Control problem-solving process structure*

Why fuzzy control?
• Ability to translate imprecise/vague knowledge of human experts
• Simple, easy to implement technology
• Software design and hardware implementation support
• Results are easy to transfer from product to product
• Smooth controller behaviour
• Robust controller behaviour
• Ability to control unstable systems

Maybe you are right. However, to investigate the problem, we constructed a mathematical model. How can we control the object otherwise?

It was Mamdani who demonstrated the way [Proc79] by constructing the first fuzzy controller. The controller was designed for a plant which comprised a steam engine and boiler combination. The model of the plant had two inputs: the heat input to the boiler and the throttle opening at the input of the engine cylinder, and two outputs: the steam pressure in the boiler and the speed of the engine. The problem in classical control was that the plant model was highly nonlinear with both magnitude and polarity of the input variables.

FUZZY LOGIC EXPLOITS THE TOLERANCE FOR IMPRECISION

High precision entails high cost	⇒ to minimise cost, minimise precision
High precision entails low solvability	⇒ to make a problem treatable, lower precision

I know it is not easy to control such a plant.

Yes, this plant possesses different characteristics at different operating points, so that the direct digital controller implemented for comparison purposes had to be tuned (by a trial and error process) to reach the best performance each time the operating point was altered.

3.2 Fuzzy rules processing

3.2.1 Mamdani-type fuzzy processing

Mamdani proposed to control the plant by realising some fuzzy rules or fuzzy conditional statements, for example:

if pressure error (PE) is **Negative Big (NB)**
then heat change (HC) is **Positive Big (PB)**

So we can measure outputs of a plant and calculate a control action according to this rules table.

What is the pressure error?

It is the difference between the current value of the pressure and the set point. And the speed error (SE) is again the difference between the current speed and the set point.

I am not a control engineer, could you explain what the set point is?

A set point is usually a desired value for the plant output: the values which some measured parameters of the process or the object should have at a particular time.

It is similar to our boat control, isn't it?

Exactly! But Mamdani proposed a modification. In order to improve the control quality, he increased the number of control inputs and used the change in pressure error (CPE), defined as the difference between the present PE and the last one (corresponding to a last sampling instant), and the change in speed error (CSE) as well.

Why does that improve the control quality?

Because it provides a controller with some degree of prediction. For example, if PE is Negative Big and CPE is also Negative Big, it means that at the next moment PE will become even more Negative Big and HC should obviously be Positive Big. But if PE is Negative Big and CPE is Positive Big, it means that at the next moment PE will become Negative Smaller and we should think about how to choose PE. This addition increases the sensitivity of the controller.

Mamdani realised his controller on the PDP-8 computer. It contained 24 rules. A fixed digital controller was also implemented on the computer and applied to the same plant for a comparison. For the fixed controller, many runs were required to tune the controller for the best performance. This tuning was done by a trial-and-error process. The quality of the fuzzy controller was found to be better than the best result of the fixed controller each time, so opening a new era in a controller design.

I do not understand why it was so important. I have heard the rule-based control was used before Mamdani.

True. But the main problem is how to process these rules. Mamdani used fuzzy theory to calculate the output according to the rules set and gained a solid theoretical base. It means that we can use this result to construct other fuzzy controllers. Some of these controllers, illustrating different possible applications, are given in Table 3.1.

Table 3.1 Fuzzy logic controllers (sample)

Date	*Application*	*Designers*
1975	Laboratory steam engine	Mamdani and Assilian
1976	Warm water plant	Kickert and Van Nauta Lemke
1977	Ship course control	Van Amerongen
1978	Rolling steel mill	Tong
1979	Iron ore sinter plant	Rutherford
1980	Cement kiln control	Umbers and King
1985	Aircraft landing control	Larkin
1989	Autonomous guided vehicle	Harris *et al.*
1991	Ship yaw control	Sutton and Jess

How are these rules processed?

In the literature, the process is called the inference mechanism or the inference engine. Let us write Mamdani rules in a general case. Mathematically a rule will look like

$$R^i: \quad \text{IF} \quad A_{i1}(x_1), A_{i2}(x_2), \ldots, A_{im}(x_m) \quad \text{THEN Y is } B_i.$$

What does it mean?

Here $x_1, x_2, \ldots x_m$ stand for input variables, for example, pressure, temperature, error, etc., A_{ij} (xj) (j = 1,2,...m) is a fuzzy set on X_j, Y is an output variable, B_i is a fuzzy set on Y.

so, in the rule:

if pressure error (PE) is ***Negative Big (NB)*** *then heat change (HC) is* ***Positive Big (PB)***

pressure error is X_1 heat change is Y, A_{i1} (x_1) is ***Negative Big****, and B_i is* ***Positive Big****.*

Right! The method is clear and easy to understand.

I understand it, but what about a computer? It is a computer that should calculate the result of this rule.

The result is fuzzy, a fuzzy set. The inputs of the inference engine are fuzzy sets and the output is also a fuzzy set. To clarify how a computer works in greater detail, let me provide you with a more formalised explanation. We will reintroduce some of the basic notions dealing with linguistic variables and properties of fuzzy sets.

3.2.2 Linguistic variables

A linguistic variable is defined by $<$ Xi, *LX*i, *X*i, A(Xi) $>$. Here Xi is the symbolic name of a linguistic variable (for example, pressure or temperature), LXi stands for the set of linguistic values that Xi can take on (**Negative Big, Positive Big** and **Zero**). The set of all possible values is *LX*. We denote an arbitrary element of *LX* by LXi. *X*i is the actual physical domain over which the meaning of the linguistic value (temperature between –50 and +100°C) is determined, *X* can be discrete or continuous. A(Xi) is a semantic function which gives a 'meaning' (interpretation) of a linguistic value in terms of the quantitative elements of X, i.e.,

$$A(X): LXi \rightarrow \mu_{LX}(x),$$

where A(X) is a denotation for a fuzzy set or a membership function defined over X, i.e.,

$$A(X) = \Sigma\, \mu_{LX}(x) / x \quad \text{in a case of discrete X,}$$
$$A(X) = \int \mu_{LX}(x) / x \quad \text{in a case of continuous X.}$$

Fuzzy rules

Most of the fuzzy logic applications involve construction and processing of fuzzy rules

Fuzzy rules serve to describe, in linguistic terms, a qualitative relationship between two or more variables

Processing of fuzzy rules or fuzzy reasoning provides a mechanism for using fuzzy rules to compute the response to a given fuzzy controller input

In other words, A(X) is a function which takes a symbol as its argument and returns the 'meaning' of this symbol in terms of the fuzzy set or the membership function of the fuzzy set.

You have not started your explanation, but I'm already lost. What is A(X)?

A(X) is a procedure which communicates the membership function for the linguistic value, for example, ***Negative Big*** for pressure (Fig. 3.2).

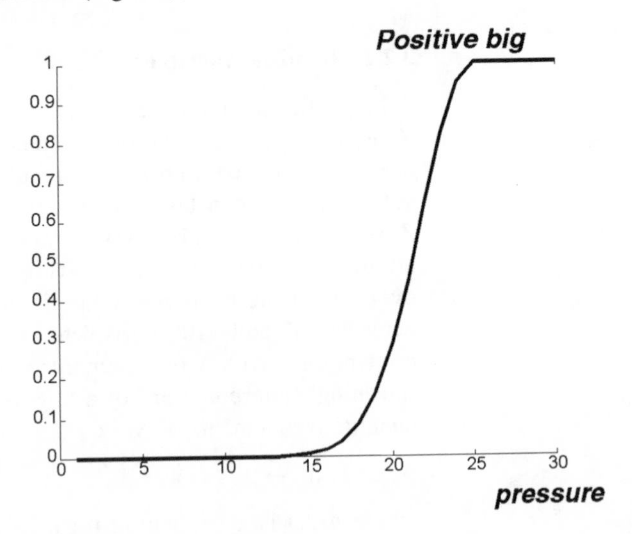

Fig. 3.2 *A membership function*

We can now define a set of m rules as

$$\text{if } X_1 \text{ is } LX_1(k) \text{ and } \ldots \text{ and } X_n \text{ is } LX_n(k) \text{ then } u \text{ is } LU(k), \quad k = 1, \ldots, m,$$

for example.

if pressure is ***high*** and temperature is ***low*** then time is ***short*** or if interest rate is ***low*** and price is ***low*** invest a ***large*** amount of money.

In a real engineering fuzzy system usually crisp inputs are used.

3.2.3 Fuzzy rules firing

What do you mean by crisp inputs?

I mean that we usually know the value of the inputs: temperature is 25°C, pressure is 2.5kPa, interest rate is 9.9%. One can apply these inputs to the fuzzy system (controller). This process of applying the inputs to the rules is called firing the rules. Actually, firing consists of computing the degree of match between the crisp input and the fuzzy sets, describing the meaning of the antecedent part of the rules, and then 'clipping' the fuzzy set describing the meaning of the consequent part of the rule, to the degree to which the antecedent part has been matched by the crisp input.

Could you explain it in greater detail, please? First of all, what are the antecedent and consequent parts?

The antecedent part is the condition part, the first part of the rule. The consequent part is the result part, the second part of the rule. For example, in the rule:

if interest rate is ***low*** < antecedent part >
and share price is ***low***

then invest a ***large*** < consequent part >
amount of money

When we apply the crisp input to the rule we compare the input value with the membership functions describing different linguistic variables (Fig. 3.3). The result of this comparison will be a set of pairs (LX, μ_{LX} (x)), e.g., if one applies the membership functions in Fig. 3.3 and the crisp value for the pressure error is –22kPa, then it will be transformed into the fuzzy input (***Negative Big***, 0.75) and (***Negative Small***, 0.05).

I understand that the fuzzy input is the result of comparing the crisp input and the membership functions assigned to describe this input. Where do these membership functions come from?

We will discuss that later in section 4.4. Moving on, the fuzzy input obtained is applied to the rules.

Which rules? All of them or just one?

One tries to apply or fire all the rules. However, just those rules where the antecedent parts contain the corresponding input, will be fired.

Example 3.1 *In the following rules table*

if pressure is ***Neg Big*** then time is ***short*** ♣ – **fired**
if pressure is ***Neg Small*** then time is ***short*** ♣ – **fired**
if pressure is ***Zero*** then time is ***average***
if pressure is ***Pos Small*** then time is ***long***
if pressure is ***Pos Big*** then time is ***long***

just first two will be fired in the situation of Fig. 3.3.

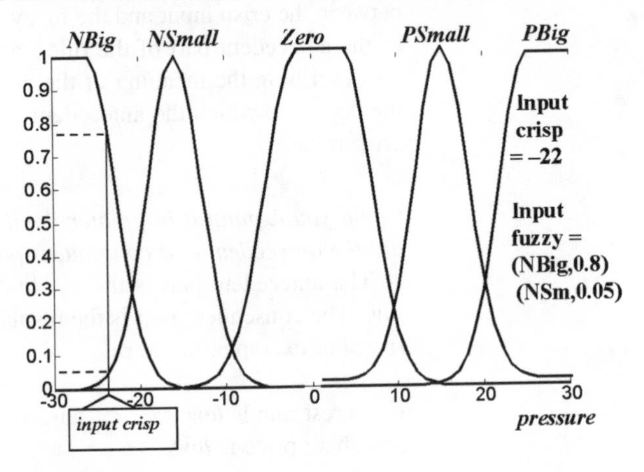

Fig. 3.3 *Fuzzification of the crisp input – pressure*

Why just these two?

Because to fire the rule, one compares the fuzzy input with the rule antecedent part. If the antecedent part contains this input, the rule is fired. Our input is Neg Small and Neg Big, so just the first two rules are fired.

What does one fire the rules for?

As a result of firing, the fuzzy input is mapped into the fuzzy output. To execute this mapping, one needs to transform the membership degree from the rule antecedent part to the rule consequent part. In our example, the output will be (***short***, 0.75), (***short***, 0.05).

*Why two **short**?*

Because that is the consequent part of two fired rules.

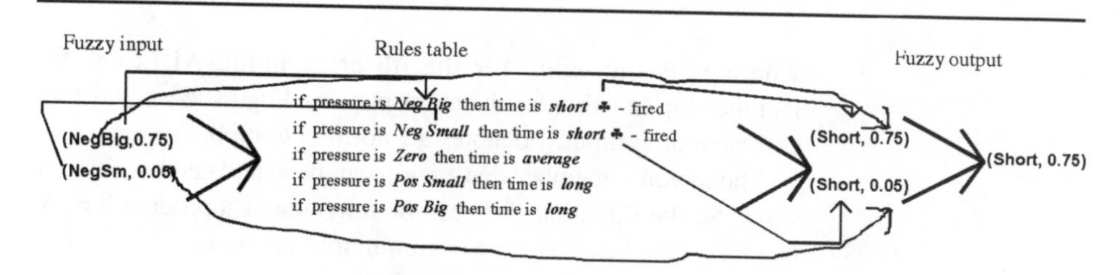

Fig. 3.4 *Fuzzy processing with the rules table*

I think we should choose one result or calculate it.

Right! As a result an $\hat{s}$-norm is usually calculated and a maximum operation is taken. So the fuzzy output will be (**short**, 0.75).

Yes, I realise what happens if we have just one condition in the antecedent part. How do we fuzzify inputs if we have two or more?

Let us again consider the rule

if pressure is ***Neg Big*** and temperature is ***high*** then time is ***short***

and the corresponding membership functions for pressure (Fig. 3.3) and temperature (Fig. 3.5). In the case when crisp inputs are pressure = –22kPa and temperature = 22°C try to fuzzify inputs.

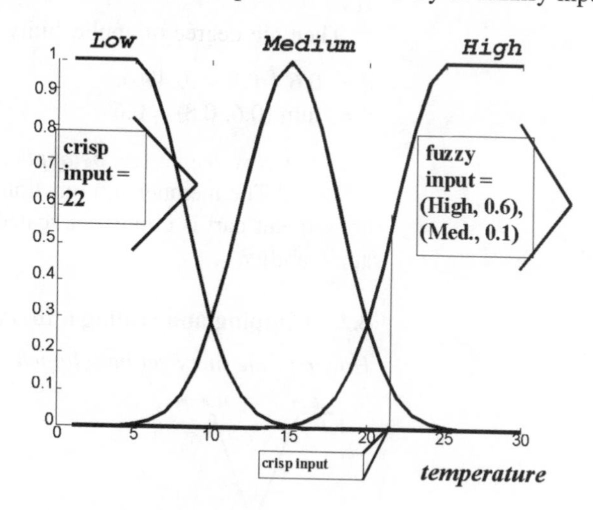

Fig. 3.5 *Fuzzification of the crisp input – temperature*

3.2.4 Calculating the applicability degree

At this stage the degree to which the whole condition part (all the inputs) satisfies the rule is calculated. This degree is called the degree of applicability of the condition part. It is denoted here as β.

$$\beta = A_1(x_1)\ {}^{t}\ A_2(x_2)\ {}^{t} \dots {}^{t}\ A_n(x_n) \tag{3.1}$$

where x_i is the value for the ith crisp input, Ai (x_i) is the membership function for the corresponding linguistic value for the corresponding input, *t* denotes a *t* norm operation.

I hope you remember that *t* norm can be calculated by different ways. So the Equation (3.1) can be rewritten as a product $\beta = A_1 (x_1) * A_2 (x_2) * \ldots * A_n (x_n)$ or minimum operation

$$\beta = \min (A_1 (x_1) , A_2 (x_2) , \ldots , A_n (x_n))$$

Hang on! I understand how the t norm can be calculated but I do not understand what corresponds to what in your explanation.

In our example we have two conditions in the antecedent part and two inputs: pressure and temperature. The membership functions for the linguistic variable pressure are given in Fig. 3.3 and the membership functions for the linguistic variable temperature are given in Fig. 3.5. Now we need to determine the degree to which the value of –22kPa for pressure matches the membership function for the linguistic value ***NegBig*** (it will be 0.75) and determine the degree to which the value of 22°C for temperature matches the membership function for the linguistic value ***high*** (it will be 0.6). Is it clear?

Yes.

Then the degree of applicability can be calculated as

$$\beta = 0.6 * 0.8 = 0.48 \text{ or}$$
$$\beta = \min (0.6, 0.8) = 0.6$$

And now this degree can be applied to the output of the rule.

Right! The membership function of the linguistic value for the consequent part is clipped or scaled to the level of the degree of applicability.

3.2.5 Clipping and scaling a fuzzy output

How can the fuzzy set be 'clipped'?

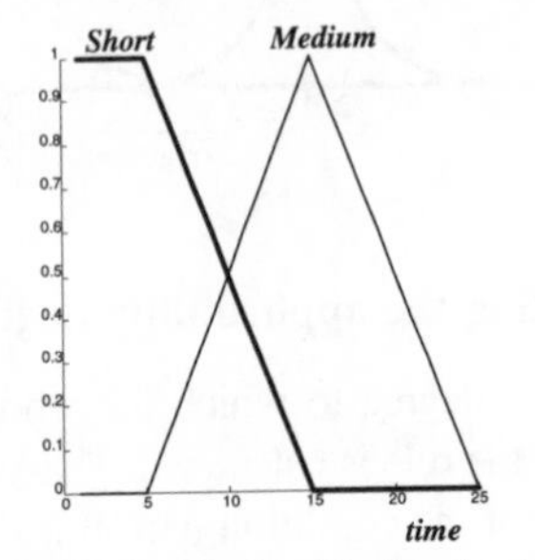

Fig. 3.6 *Membership function for the rule consequent part*

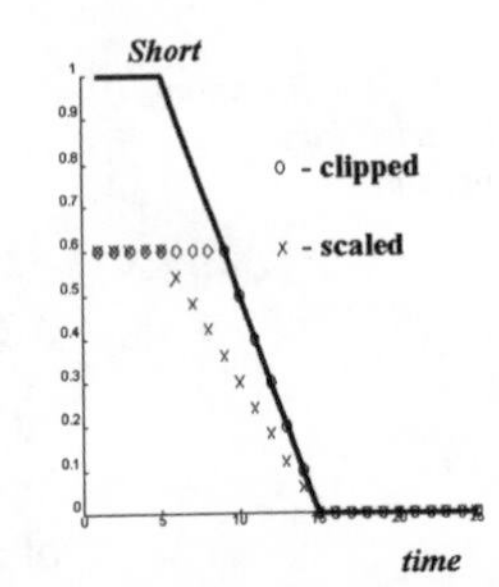

Fig. 3.7 *Scaled and clipped membership function*

This process is illustrated in Fig. 3.7. Let us denote the 'clipped' fuzzy set as CLU. The height of CLU(k) is equal to the degree of match of the kth rule antecedent degree of applicability, and is denoted by βk. The peak value of CLU(k) is equal βk. Actually here we cut the membership function of the unclipped version LU(k) on the level βk. In our example we need to clip the membership function of the linguistic value **short** for the output time on the level of 0.48 or 0.6 .

How do you scale the fuzzy set?

You need to remember that even when we discuss clipping and scaling fuzzy sets, we actually talk about their membership functions. So I reformulate your question as: how do you scale the membership function? This method can be applied if you think clipping is not fair enough.

Not fair enough?

Yes. In the case of clipping, one assigns the same degree to a range of the input values regardless of the membership degrees of the unclipped membership functions. In the case of scaling, you change the membership degrees for all the elements of the universe. A scale rate depends on an applicability degree. A scaled membership function should have the peak value equal to the applicability degree. It means that in order to produce the scaled membership function, we need to multiply all the membership degrees by the value of β. It is only valid in a case when a peak value of the unclipped fuzzy set was 1. If it was α, then the multiplication factor should be β/a.

What will one have as a result of a rule application?

Each rule produces the clipped or scaled fuzzy set, or more exactly, the clipped or scaled membership function.

This is not a result of fuzzy processing, is it?

What do you think?

I guess it cannot be the result, because until now we have discussed just processing one rule and usually we have at least a few.

Right! A fuzzy system includes some rules. Each rule produces the fuzzy set. To obtain the result of the whole set of rules processing or the fuzzy output of the system, one needs to combine the resulting fuzzy sets from all the rules.

I know we may determine the final result as the union of all the rules results

$$U = U_1 \cup U_2 \cup \ldots \cup U_m$$

where U_i is the result of applying the ith rule and m is a number of rules applied.

That's it! Finally, the overall control output U is obtained as the union of the clipped or scaled control outputs. U may be a convex fuzzy set or may be a non-convex fuzzy set which consists of a number of convex fuzzy subsets. You should remember that U_i is a clipped fuzzy set CLU_i or a scaled fuzzy set SLU_i. Do you remember how to calculate the union?

There are different ways. For example the maximum operation can be used

$$\mu_U = max\,(\mu_{U1}, \mu_{U2}, \ldots \mu_{Um})$$

or one can apply any other ŝ-norm operation.

3.2.6 Sugeno-type fuzzy processing

We have considered how fuzzy processing was realised by Mamdani, and the method is called after him. Another method was proposed by Sugeno, who changed a part of the rules. In his method, the consequent part is just a mathematical function of the input variables. The format of the rule is:

if $A_1(x_1), A_2(x_2), \ldots, A_n(x_n)$ then $Y = f(x_1, x_2, \ldots, x_n)$.

You see that an antecedent part is similar to the Mamdani method. The function f in a consequent is usually a simple mathematical function, linear or quadratic:

$$f = a_0 + a_1 \times x_1 + a_2 \times x_2 + \ldots + a_n \times x_n$$

The antecedent part in this case is processed in exactly the same way as the Mamdani method, and then an obtained degree of applicability is assigned to the value of Y calculated as the function of real inputs.

Can you give an example.

Let us again consider the rule:

if pressure is ***NegBig*** and temperature is ***high*** then time is ***short*** but now replace it with:

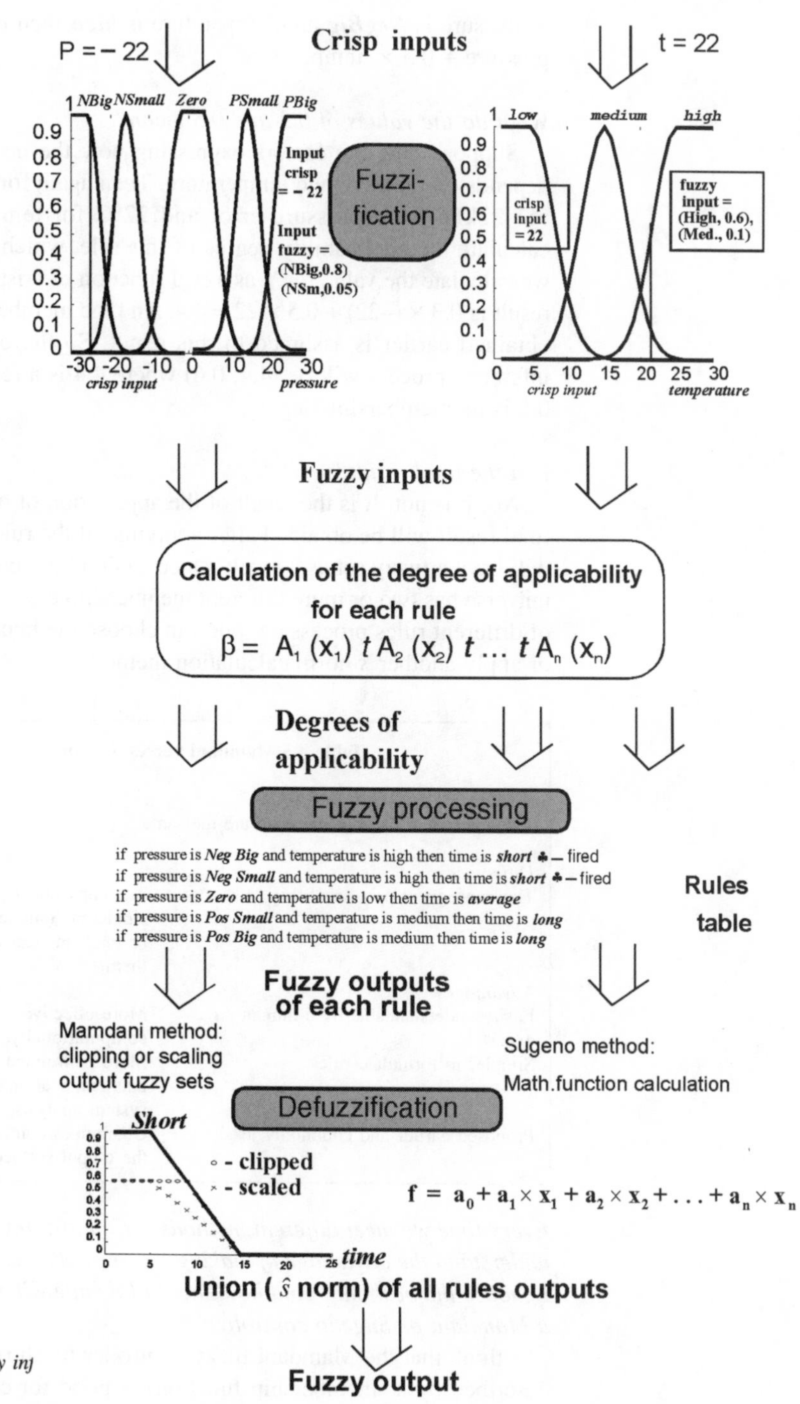

Fig. 3.8 *Fuzzy inj* *structure*

if pressure is ***NegBig*** and temperature is ***high*** then time is 0.3 × pressure + 0.5 × temperature.

What do the values of 0.3 and 0.5 mean?

Suppose they are factors expressing how the necessary time depends on pressure and temperature. Then again for crisp inputs of –22kPa for a pressure error and 22°C for temperature we calculate the applicability degree of this rule, which is 0.6. Now we calculate the value for Y as a real function of crisp inputs. The result is 0.3 × (–22) + 0.5 × 22 = 4.4. And the membership degree obtained earlier is assigned to this result. So the output of the inference process will be (4.4, 0.6) where 4.4 is a real result and 0.6 is its membership degree.

Is it the final result?

No, it is not. It is the result of the application of one rule. The final result will be obtained after applying all the rules. Then one will have a fuzzy set as a result. Once again if any element of the universe has two or more different membership degrees as a result of different rules processing, one can choose the maximum value or apply another $\hat{s}$ norm calculation method.

Table 3.3 Mamdani verses Sugeno

Mamdani	Sugeno
Similarity:	
The antecedent parts of the rules are the same.	
Difference:	
The consequent parts are fuzzy sets.	The consequent parts are singletons (single spikes) or mathematical functions of them.
Advantages:	
Easily understandable by a human expert.	More effective computationally.
Simpler to formulate rules.	More convenient in mathematical analysis and in system analysis.
Proposed earlier and commonly used.	Guarantees continuity of the output surface.

Every time we meet different methods for almost any operation. I understand the advantage of making my own choice, but how can I choose the best one? And how should I decide whether to apply a Mamdani or Sugeno controller?

I think that the Mamdani fuzzy controller (each rule output is described by a membership function) is good for capturing the expertise of a human operator. But it is awkward to design if you have the plant model but don't have a working controller (for instance, a human operator). Sugeno fuzzy controller (each rule's output is a linear equation) is good for embedding linear controller and continuous switching between these output equations. (In fact, gain scheduling!). This becomes very effective when the plant model is known. Also an adaptive capability and mathematical tractability make this type of fuzzy controller a primary choice for nonlinear and/or adaptive control design that is subject to rigorous analysis.

To summarise our discussion about fuzzy inference, we should say that this procedure can be roughly divided into three steps:

- Calculation of the degree of applicability of the antecedent (condition) part of each control rule.
- Calculation of the inference result (solution fuzzy set) for each control rule.
- Synthesis of the solution set for each control rule into the fuzzy output set.

This process is illustrated on Fig. 3.8.

I think fuzzy inference is very complicated. It contains too many operations and not one of them is easy to understand. Most of these operations are nonlinear mappings.

Maybe you are right. However, this complication allows us to create complicated input/output characteristics with a very small number of control rules. The designer should formulate a small number of simple rules, the computer should perform rather complicated calculations. Isn't that fair enough?

3.3 Fuzzy controller operation

The inference engine is the heart of a fuzzy controller (and any fuzzy rules system) operation. Its actual operation can be divided into three steps (Fig. 3.9):

- Fuzzification – actual inputs are fuzzified and fuzzy inputs are obtained.
- Fuzzy processing – processing fuzzy inputs according to the rules set and producing fuzzy outputs.
- Defuzzification – producing a crisp real value for a fuzzy output.

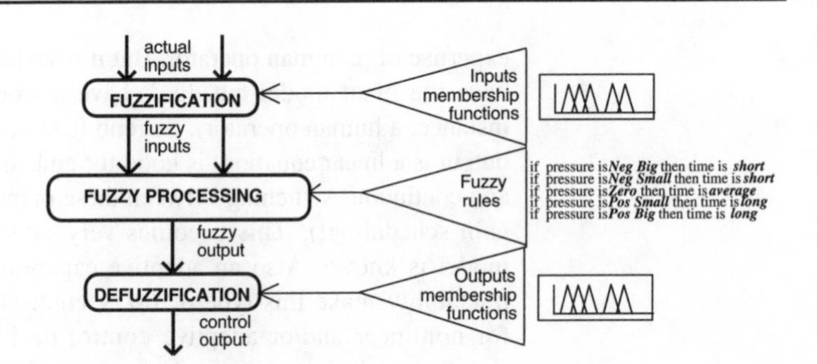

Fig. 3.9 *Operation of a fuzzy controller*

We have considered the first two stages. What do we need the third one for?

In a real control system, the controller output should be used to control a real object or process. So we need to know a crisp value for every output signal. Defuzzification produces this value on the basis of output membership functions.

Oh! We have spent a lot of time studying this fuzzy inference in order to come back to the crisp output. What did we need this fuzzy stuff for?

The main reason is that fuzzy control gives us a rather simple to use method for producing high quality controllers with complicated input/output characteristics.

Can you describe this defuzzification process.

Yes, it is not very difficult. But once again a few methods have been developed and to help you to solve the choice problem, we consider this process later in Section 4.6 in greater detail.

3.4 Structure of a simple open-loop fuzzy controller

After you have considered how a fuzzy controller operates, we can discuss how to design simple fuzzy controllers.

I understand that in order to construct a fuzzy controller, one needs just to write some rules.

It's not so simple! The classical design scheme contains the following steps:

1. Define the input and control variables – determine which states of the process shall be observed and which control actions are to be considered.
2. Define the condition interface – fix the way in which observations of the process are expressed as fuzzy sets.

3. Design the rule base – determine which rules are to be applied under which conditions.
4. Design the computational unit – supply algorithms to perform fuzzy computations. Those will generally lead to fuzzy outputs.
5. Determine rules according to which fuzzy control statements can be transformed into crisp control actions.

The typical structure of a fuzzy controller is given in Fig. 3.10.

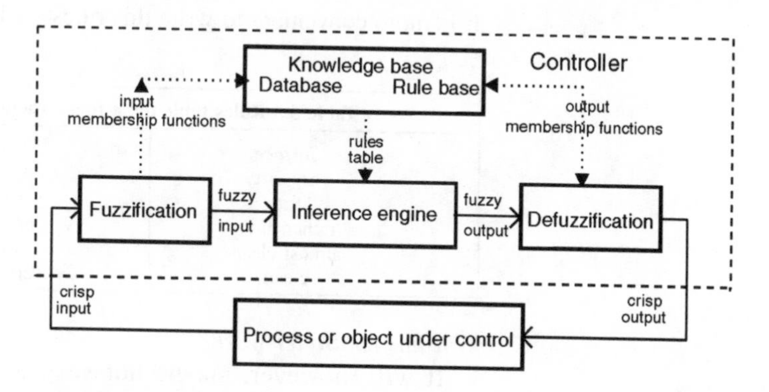

Fig. 3.10 *The fuzzy logic controller (a basic structure)*

There seem to be so many problems! How can we solve them?

Step by step! The first one is a usual step in the design of any controller. You choose variables which can be measured. They become the inputs of the controller. Step 2 represents the fuzzification process, Step 4 fuzzy inference and Step 5 defuzzification process.

However, you are right, the heart of a fuzzy controller design is a formulation of the rules.

Where can I get these rules from?

The main basis is an expert's experience, his/her understanding how a fuzzy controller should operate and what it should do.

Example 3.2 *A fuzzy vacuum cleaner*

Let us try to develop the rules table for the fuzzy controller of a vacuum cleaner. This controller should regulate the force of sucking dust from a surface being cleaned. This force can be described as a linguistic variable with values: *very strong, strong, ordinary, weak, very weak*. The input of this controller should obviously consider an amount of dust on the surface. The surface can be *very dirty, dirty, rather dirty, almost clean, clean*. The controller can change the force depending on how dirty the surface is. One can propose the following set of rules to describe the controller operation:

if surface is ***very dirty*** **then** force is ***very strong***,
if surface is ***dirty*** **then** force is ***strong***,
if surface is ***rather dirty*** **then** force is ***ordinary***,
if surface is ***almost clean*** **then** force is ***weak***,
if surface is ***clean*** **then** force is ***very weak.***

It is more convenient to write this rules set in a table form.

Table 3.3 Rules table for a fuzzy vacuum cleaner

Surface	*Force*
very dirty	very strong
dirty	strong
rather dirty	ordinary
almost clean	weak
clean	very weak

Will this controller work?

It will, however, maybe not very well. To improve the performance, one should apply some extra expert's knowledge. So perhaps a driving force should depend not only on an amount of dust, but on the surface texture and fabric also. There is some difference in cleaning wood and wool, for example. So let us introduce another input: surface type with linguistic values of wood, tatami, carpet. Then we can implement the following rules table.

Table 3.4 Rules table for surface type and dust amount

	clean	almost clean	rather dirty	dirty	very dirty
wood	*very weak*	*very weak*	*weak*	*ordinary*	*strong*
tatami	*very weak*	*weak*	*ordinary*	*strong*	*very strong*
carpet	*weak*	*ordinary*	*ordinary*	*strong*	*very strong*

Could this controller now be used?

Actually a controller like this has been applied in a Japanese manufactured vacuum cleaner [Hir93]. A block diagram of the controller is given in Fig. 3.11. You see that the fuzzy controller has two inputs and one output.

I do not understand how this controller can work in a case of, say, a linen fabric surface. Can it be applied for cleaning wood, tatami and carpets only?

These names, or linguistic labels, have a symbolic sense only. They just mark different membership functions which should describe how easy it is to clean a particular surface, i.e. to mark a different degree of 'easyness'. And remember these degrees are fuzzy, in that a fabric can be considered as 'a little tatami and mainly wood'.

And what about linen?

This could be placed between tatami and carpet and considered as a little of this and a little of that.

How is this vacuum cleaner implemented practically?

The dust sensor includes a photo transistor which is mounted opposite an infrared light-emitting diode. Infrared rays are emitted in a beam. When they pass through the dust, some rays are lost, causing the amount which reaches the photo transistor to decrease. The varying component is amplified and used to evaluate the amount of dust on the surface being cleaned.

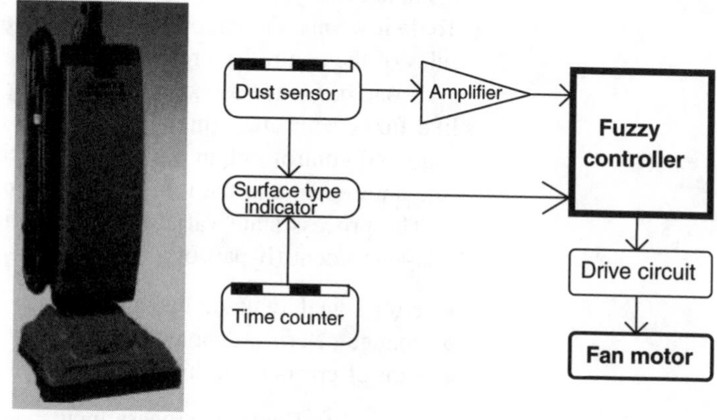

Fig. 3.11 *The structure of the vacuum cleaner fuzzy controller*

What about the second input? How is the surface type evaluated?

As cleaning proceeds, the amount of dust decreases, but the speed of decreasing depends on the surface type. If the surface is smooth like wood, it is cleaned very fast because it is easy to pick up dust from such a surface. On the other hand, it is hard to clean wool carpet surfaces. Thus, by evaluation of the change of the dust amount collected during a time unit, we can judge the type of the surface being cleaned. This controller works well under different conditions.

So in this controller, the first input is the amount of dust collected during a time unit. And the second input is the change in this amount. Is that right?

Yes, it is.

Then the second input can be considered as a derivative of the first one. Is that right?

Yes, it is.

So our fuzzy controller is an analogy to a classical PD-controller, isn't it?

Yes, you are right.

3.5 Structure of a feedback PID-like fuzzy controller

3.5.1 Fuzzy controllers as a part of a feedback system

We were taught how to design conventional PID controllers. How can I design a fuzzy PID controller?

It is not easy to design an exact PID controller and I think it is not necessary, but one can construct a PID-like fuzzy controller. To do it we need to choose the input and output variables and the rules of the controller properly.

If one has made a choice of designing a P-, PD-, PI- or PID-like fuzzy controller, this already implies the choice of process state and control output variables, as well as the content of the rule-antecedent and the rule-consequent parts for each rule.

The process state variables representing the contents of the rule-antecedent (**if**-part of a rule) are selected among:

- error signal, denoted by e;
- change-of-error, denoted by Δe;
- sum-of-errors or an integral error, denoted by Σe.

The control output (process input) variables representing the contents of the rule-consequent (**then**-part of the rule) are selected among:

- change-of-control output, denoted by Δu;
- control output, denoted by u.

Hang on! What is this error?

The error is the difference between the desired output of the object or process under control or the set-point and the actual output.

This is one of the basic milestones in conventional feedback control. Furthermore, by analogy with a conventional controller, we have:

- $e(t) = y_{sp} - y(t);$
- $\Delta e\,(t) = e(t) - e(t-1);$
- $\Delta u\,(t) = u(t) - u(t-1).$

In the above expressions, y_{sp} stands for the desired process output or the set-point, y is the process output variable (control variable); k determines the current time.

By the way, considering the classes of a PID-controller, what type of a controller can be represented by the boat controller designed in the introduction?

I think the deviation can be considered as an error and the turn as a control output. Then this table represents a choice of the control output roughly proportional to the error. It means that we have designed a P-like controller.

That's right! Now let us try to design the PD-like controller.

3.5.2 PD-like fuzzy controller

The equation giving a conventional PD-controller is

$$u(k) = K_P \times e(t) + K_D \times \Delta e(t),$$

where K_P and K_D are the proportional and the differential gain factors.

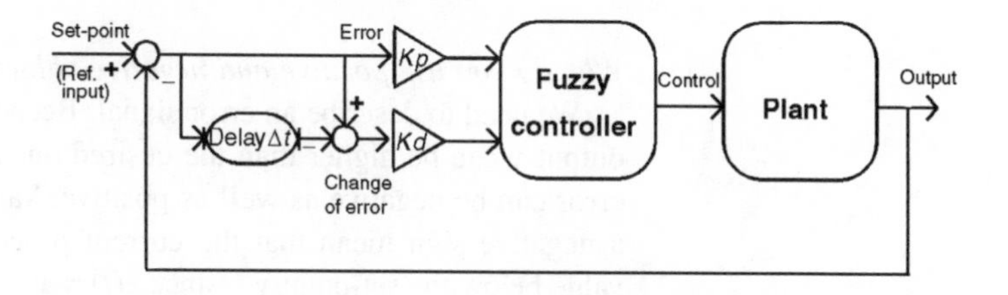

Fig. 3.12 *A block-diagram of a PD-like fuzzy control system*

If I try to describe this equation with the help of rules, what inputs and outputs should I use for the rules table?

I do not know.

Let us consider this equation. The PD controller for any pair of the values of error (e) and change-of-error (Δe) calculates the control signal (u). The fuzzy controller should do the same thing. For any pair of error and change-of-error, it should work out the control signal. Then a PD-like fuzzy controller consists of rules, and a symbolic description of each rule is given as

if $e(t)$ is <property symbol> and $\Delta e(t)$ is <property symbol> then $u(t)$ is <property symbol>, where <property symbol> is the symbolic name of a linguistic value.

The natural language equivalent of the above symbolic description reads as follows. For each sampling time. t:

if the value of error is <linguistic value> and the value of change-of-error is <linguistic value> then the value of control output is <linguistic value>.

We will omit the explicit reference to sampling time t, since such a rule expresses a casual relationship between the process state and control output variables, which holds for any sampling time t.

What does the symbolic name of a linguistic value mean?

This is one of the linguistic qualifiers, determined for the proper variable: error, change-of-error or control signal, for example: ***high, low, medium***, etc.

So I have to have membership functions, describing all these qualifiers for all my variables: error, change-of-error or control.

Definitely. And you should remember that these variables may be measured in different units. So the rules of the PD controller can be like:

if error is ***Positive Big*** and change-of-error is ***Negative Big*** then control is ***Negative Small***

Why do you use positive and negative values?

We need to describe an error signal. Because the actual process output y can be higher than the desired one as well as lower, the error can be negative as well as positive. Values of error (e) with a negative sign mean that the current process output $y(t)$ has a value below the set-point y_{sp} since $e(t) = y_{sp} - y(t) < 0$. A negative value describes the magnitude of the difference $y_{sp} - y$. On the other hand, linguistic values of e with a positive sign mean that the current value of y is above the set-point. The magnitude of such a positive value is the magnitude of the difference $y_{sp} - y$.

The change-of-error (Δe) with a negative sign means that the current process output $y(t)$ has increased when compared with its previous value $y(t-1)$, since $\Delta e(t) = e(t) - e(t-1) = -y(t) + y(t-1) < 0$. The magnitude of this negative value is given by the magnitude of this increase. Linguistic values of $\Delta e(t)$ with a positive sign mean that $y(t)$ has decreased its value when compared to $y(t-1)$. The magnitude of this value is the magnitude of the decrease.

Linguistic values of e with a negative sign mean that the current process output y has a value below the set-point y_{sp} since $e(t) = y_{sp} - y(t) < 0$. The magnitude of a negative value describes the magnitude of the difference $y_{sp} - y$. On the other hand, linguistic values of e with a positive sign mean that the current value of y is above the set-point. The magnitude of such a positive value is the magnitude of the difference $y_{sp} - y$.

Linguistic values of Δe with a negative sign mean that the current process output $y(t)$ has increased when compared with its previous value $y(t-1)$ *since* $\Delta e(t) = -(y(t) - y(t-1)) < 0$. The magnitude of such a negative value is given by the magnitude of this increase. Linguistic values of $\Delta e(t)$ with a positive sign mean that $y(t)$ has decreased its value when compared to $y(t-1)$. The magnitude of such a value is the magnitude of the decrease.

A linguistic value of 'zero' for e means that the current process output is about the set-point. A 'zero' for Δe means that the current process output has not changed significantly from its previous value, i.e. $-(y(t) - y(t-1)) = 0$. The sign and the magnitude for u constitutes the value of the control signal.

3.5.3 Rules table notation

I want to propose to you a convenient form to write down rules. This table form is suitable when we have two inputs and one output. On the top side of the table we should write the possible linguistic values for the change-of-error (Δe) and on the left side, the error (e). The cell of the table at the intersection of the row and the column will contain the linguistic value for the output corresponding to the value of the first input written at the beginning of the row and to the value of the second input written on the top of the column.

Let us consider the example [Drian93] where both inputs and an output have a set of possible linguistic values {NB, NM, NS, Z, PS, PM, PB} where NB stands for **Negative Big**, NM stands for **Negative Medium,** NS stands for **Negative Small**, Z stands for **Zero**, PS stands for **Positive Small,** PM stands for **Positive Medium** and PB stands for **Positive Big** (Table 3.7).

The cell defined by the intersection of the first row and the first column represents a rule such as:

if $e(t)$ is NB and $\Delta e(t)$ is NB **then** $u(t)$ is NB

Table 3.5

e \ Δe	PB	PM	PS	Z	NS	NM	NB
PB	NB	NB	NB	NB	NM	NS	Z
PM	NB	NB	NB	NM	NS	Z	PS
PS	NB	NB	NM	NS	Z	PS	PM
Z	NB	NM	NS	Z	PS	PM	PB
NS	NM	NS	Z	PS	PM	PB	PB
NM	NS	Z	PS	PM	PB	PB	PB
NB	Z	PS	PM	PB	PB	PB	PB

Legend
Gr. 0
Gr. 1
Gr. 2
Gr. 3
Gr. 4

How do we fill in this table?

This table includes 49 rules. We are taking into account now not just the error but the change-of-error as well. It allows to describe the dynamics of the controller. To explain how this rules set works and how to choose the rules, let us divide the set of all rules into the following five groups:

Group 0: In this group of rules both e and Δe are (positive or negative) small or zero. This means that the current value of the process output variable y has deviated from the desired level (the set-point) but is still close to it. Because of this closeness the control signal should be zero or small in magnitude and is intended to correct small deviations from the set-point. Therefore, the rules in this group are related to the steady-state behaviour of the process. The change-of-error, when it is **Negative Small** or **Positive Small,** shifts the output to negative or positive region, because in this case, for example, when $e(t)$ and $\Delta e(t)$ are both **Negative Small** the error is already negative and, due to the negative change-of-error, tends to become more negative. To prevent this trend, one needs to increase the magnitude of the control output.

Group 1: For this group of rules $e(t)$ is **Positive Big** or **Medium** which implies that $y(t)$ is significantly above the set-point. At the same time since $\Delta e(t)$ is negative, this means that y is moving towards the set-point. The control signal is intended to either speed up or slow down the approach to the set-point. For example, if $y(t)$ is much below the set-point ($e(t)$ is **Positive Big**) and it is moving towards the set-point with a small step ($\Delta e(t)$ is **Negative Small**) then the magnitude of this step has to be significantly increased ($u(t)$ is **Negative Medium**). However, when $y(t)$ is still much below the set-point ($e(t)$ is **Positive Big**) but it is moving towards the set-point very fast ($\Delta e(t)$ is **Negative Big**) no control action can be recommended because the error will be compensated due to the current trend.

Group 2: For this group of rules $y(t)$ is either close to the set-point ($e(t)$ is **Positive Small, Zero, Negative Small**) or significantly above it (**Negative Medium, Negative Big**). At the same time, since $\Delta e(t)$ is negative, $y(t)$ is moving away from the set-point. The control here is intended to reverse this trend and make $y(t)$, instead of moving away from the set-point, start moving towards it. So here the main reason for the control action choice is not just the current error but the trend in its change.

Group 3: For this group of rules $e(t)$ is **Negative Medium** or **Big,** which means that $y(t)$ is significantly below the set-point. At the same time, since $\Delta e(t)$ is positive, $y(t)$ is moving towards the set-point. The control is intended to either speed up or slow down the approach to the set-point. For example, if $y(t)$ is much above the set-point ($e(t)$ is **Negative Big**) and it is moving towards the set-point with a somewhat large step ($\Delta e(t)$ is **Positive Medium**), then the magnitude of this step has to be only slightly enlarged ($u(t)$ is **Negative Small**)

Group 4: The situation here is similar to the Group 2 in some sense. For this group of rules $e(t)$ is either close to the set-point (**Positive Small, Zero, Negative Small**) or significantly above it (**Positive Medium, Positive Big**). At the same time since $\Delta e(t)$ is positive $y(t)$ is moving away from the set-point. This control signal is intended to reverse this trend and make $y(t)$ instead of moving away from the set-point start moving towards it.

So to design a PD-like controller we need just to create a rules table like Table 3.5.

Not always. The contents of the table can be different. For example, you may replace the rule:

if e is PS and Δe is PM **then** u is NB

with the rule:

if e is PS and Δe is PM **then** u is NM.

We will discuss the concrete choice of the rules in greater detail in Section 4.5.

3.5.4 PI-like fuzzy controller

The equation giving a conventional PI-controller is

$$u(t) = K_P \times e(t) + K_I \times \int e(t)\,dt, \qquad (3.2)$$

where K_P and K_I are the proportional and the integral gain coefficients. A block diagram for a fuzzy control system looks like Fig. 3.13.

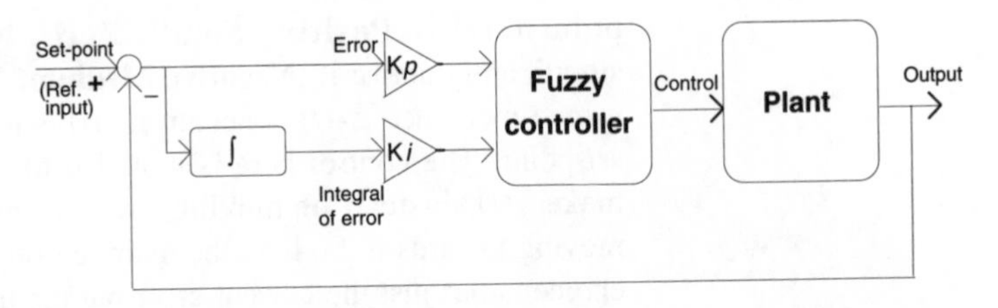

Fig. 3.13 *A block-diagram of a PI fuzzy control system (version 1)*

I see this diagram has a different form from the previous one. We have replaced differentiation with integration and a change-of-error with an integral error. How do we formulate the rules table?

Now the fuzzy controller and the rules table have other inputs. It means that the rules themselves should be reformulated. Sometimes it is difficult to formulate rules depending on an integral error, because it may have the very wide universe of discourse.

What can we do in this case?

We can move the integration from the part preceding to a fuzzy controller to the part following it. We can integrate the output of a controller, not the input. Then we may have the error and the change of error inputs and still realise the PI-control.

How does it work?

When the derivative, with respect to time, of the Equation (3.2) is taken, it is transformed into an equivalent expression

$$du(t) / dt = K_P \times de(t)/dt + K_I \times e(t)$$

or in the discrete form

$$\Delta u(t) = K_P \times \Delta e(t) + K_I \times e(t)$$

One can see here that one has the error and the change-of-error inputs and one needs just to integrate the output of a controller. One may consider the controller output not as a control signal, but as a change in the control signal. The block diagram for this system is given in Fig. 3.14. You should remember, that the gain factor K*i* is used with the error input and K*p* with the change-of-error.

Can you propose the rule format for this controller?

We can write the rule as:

if *e is <property symbol>* ***and*** *Δe is <property symbol>*
then *Δu is <property symbol>.*

In this case, to obtain the value of the control output variable $u(t)$, the change-of-control output $\Delta u(t)$ is added to $u(t-1)$. It is necessary to stress here that this takes place outside the PI-like fuzzy controller, and is not reflected in the rules themselves.

Now try to compile the rules table for this type of a controller.

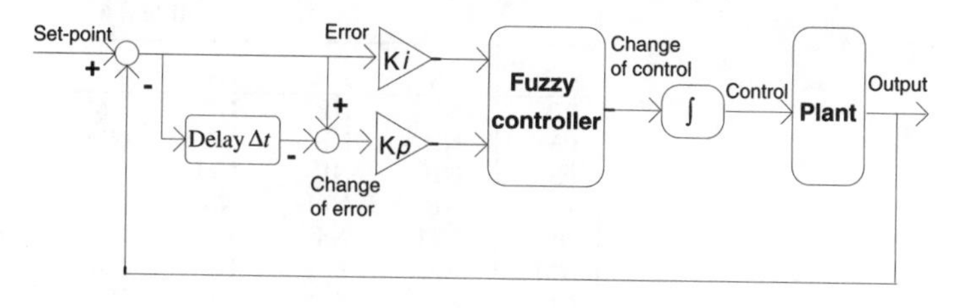

Fig. 3.14 *A block-diagram of a PI fuzzy control system (version 2)*

Can we use the table similar to the previous one?

Yes, you can. But you should remember that now the output is not a control signal but the change-of-control.

I'm not going to use the same table, let me change something.

Table 3.5

e \ Δe	**PB**	**PM**	**PS**	**Z**	**NS**	**NM**	**NB**
PB	NB	NB	NB	NB	NM	NS	Z
PM	NB	NB	NB	NM	NS	Z	PS
PS	NB	NB	NM	NS	Z	PS	PM
Z	NB	NM	NM	Z	PM	PM	PB
NS	NM	NS	Z	PS	PM	PB	PB
NM	NS	Z	PS	PM	PB	PB	PB
NB	Z	PS	PM	PB	PB	PB	PB

We have changed two rules:

if e is ***Z*** *and* Δe *is* ***NS*** *then* Δu *is* ***PM***
and if e is ***Z*** *and* Δe *is* ***PS*** *then* Δu *is* ***NM.***

Will our controller become better?

Your correction will lead to the change of the control surface. Your controller will become more reactive in the neighbourhood of the set-point. It means that even small deviation errors will be followed by larger control signals. It is difficult to say if it will become better in a general case. However, usually a designer tries to make the control surface smoother in the vicinity of a set-point. So I do not think your modifications have been very successful.

OK. If we want to make our controller less reactive to the large errors, what should we do?

Positive large or negative large?

Both.

In this case you need to modify the top and the bottom rows, maybe like this:

Table 3.6

e \ Δe	PB	PM	PS	Z	NS	NM	NB
PB	NB	NB	NB	NM	NS	Z	Z
PM	NB	NB	NB	NM	NS	Z	PS
PS	NB	NB	NM	NS	Z	PS	PM
Z	NB	NM	NS	Z	PS	PM	PB
NS	NM	NS	Z	PS	PM	PB	PB
NM	NS	Z	PS	PM	PB	PB	PB
NB	Z	Z	PS	PM	PB	PB	PB

If it is not enough, can we change these rows more significantly?

You can do whatever you want. However, if I change the left bottom corner to PS, for example, there will be a gap between two adjacent cells. So when e is changed a little bit from PM to PB, the output will jump from NS to PS. Generally speaking you should avoid these gaps and try to perform a smooth transformation between adjacent cells (see Section 4.5 for details).

It means that if we want to make significant modifications, it is best to make changes to the regions than to changes in individual cells.

You are absolutely right!

3.5.5 PID-like fuzzy controller

The equation for a PID-controller is as follows:

$$u = K_p \times e + K_d \times e + K_i \times \int edt.$$

Thus, in the discrete case of a PID-like fuzzy controller one has an additional process state variable, namely sum-of-errors, denoted σe and computed as:

$$\sigma e(t) = \sum_{i=1}^{t} e(i).$$

Have a look at the rules for PD- and PI-like fuzzy controllers and try to write down the rules format for a PID-like controller.

The symbolic expression for a rule of a PID-like fuzzy controller is:

if e is < property symbol > and Δe is < property symbol > and *σe is < property symbol> then u is < property symbol >*

Good! What is the main difference between the rules for these controllers?

The last one has three conditions in the antecedent part but the previous ones had just two. So we will need to formulate many more rules to describe the PID-controller.

Why?

If any input is described with seven linguistic values, as it was before, then because the PID-controller has three inputs and any rule has three conditions we will need 7 × 7 × 7 = 343 rules. Previously we had just 7 × 7 = 49 rules.

It is too much work to write 343 rules. The PID-like fuzzy controller can be constructed as a parallel structure of a PD-like fuzzy controller and a PI-like fuzzy controller (Fig. 3.15) with the output approximated as:

$$u = (Kp/2 \times e + Kd \times de/dt) + (Kp/2 \times e + Ki \times \int e dt).$$

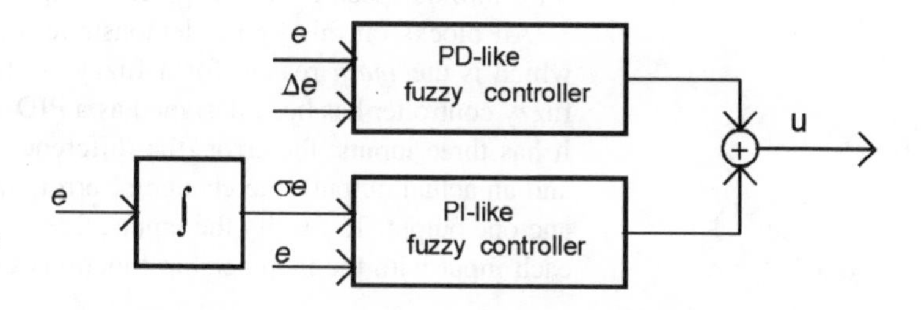

Fig. 3.15 *The structure for a PID-like fuzzy controller*

Should we always use these inputs? An error and a change of error?

Of course not. When information about the object or process under control and its structure is available, one may not want to be confined to using error, change of error, and sum of errors as process state variables, but rather use the actual process state variables. The symbolic expression for a rule in the case of multiple inputs and a single output (MISO) system is as follows:

if x_1 is <property symbol> **and** ... **and** x_n is <property symbol> **then** u is <property symbol>

Rules of this **if – then** type are usually derived from a fuzzy process model.

Example 3.3 *The fuzzy controller for steam turbines [Kiup94]*

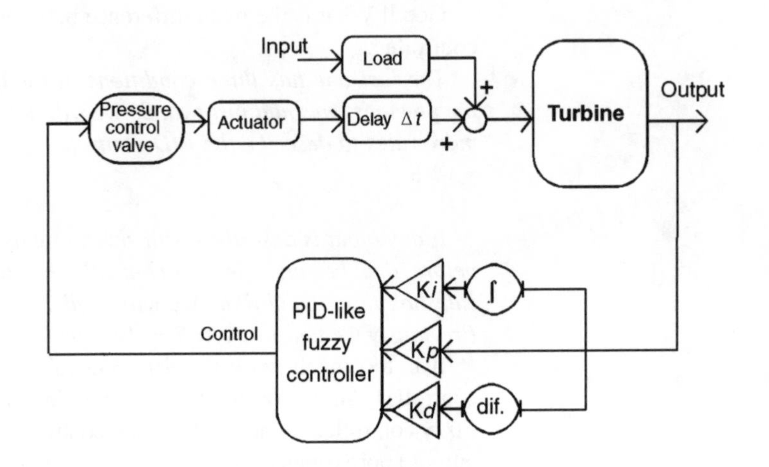

Fig. 3.16 *The block diagram of a fuzzy control system for a turbine speed control [Kiup94]*

The fuzzy controller has been designed to control the turbine speed and pressure. The block diagram of a fuzzy control system for a turbine speed control is given in Fig 3.16.

All blocks on this figure demonstrate a nonlinear behaviour, which is the *main* reason for a fuzzy control application. The fuzzy controller has been designed as a PID-like fuzzy controller. It has three inputs: the error (the difference between a set-point and an actual output), the change of error, and the integral-error, and one output. To fuzzify the inputs, three classes are applied for each input with the membership functions given in Fig. 3.17a.

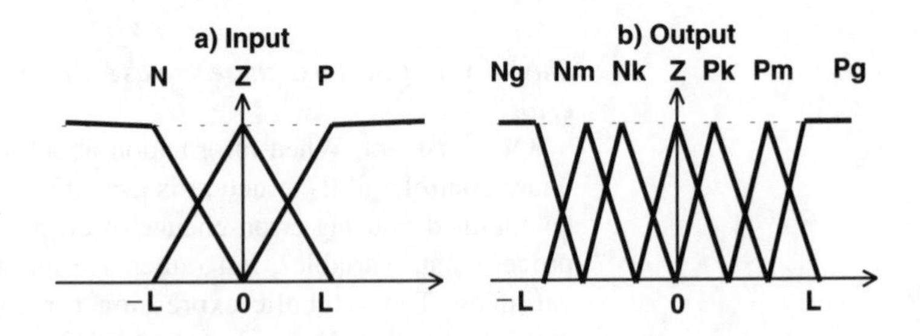

Fig. 3.17 *Membership functions for the inputs and outputs*

In order to defuzzify the output, seven classes are applied with the membership functions presented in Fig. 3.17b. Because the fuzzy controller has three inputs, its rules table has a three-dimensional image given in Fig. 3.18.

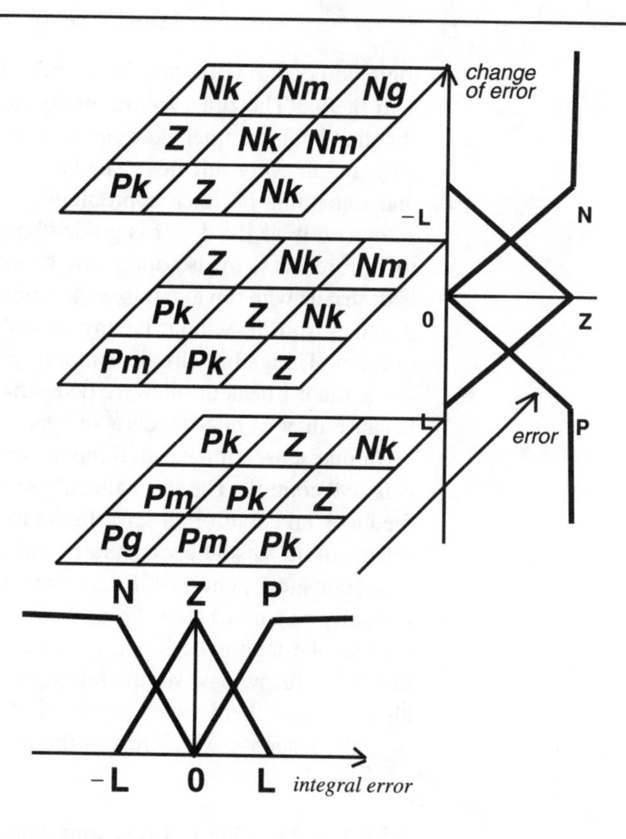

Fig. 3.18. *The rules table [Kiup94] for the fuzzy controller*

Example 3.4

I recently had to design a fuzzy-controller for backing up a truck and trailer combination. I went all through fuzzy theory, but found the best way to improve performance was to increase the number of rules used. Whether they were fuzzy or classical didn't really matter; but the non-fuzzy rules were much easier to implement. So, is fuzzy logic just a phrase the Japanese put on their consumer electronics to help it sell? What real difference does it make?

Some type of servo operators are much simpler with standard rules (e.g., 'if the sensor goes past this line, then actuate'). A truck and trailer, however, might have to recognise moving objects (e.g., birds or people), determine if some objects require a wider margin (e.g., walls or other trucks), or might need to work in less than optimal conditions (e.g., dark, fog, diesel smoke) or work with inefficient sensors. In all of these cases, Boolean rules fail due either to complexity or special cases.

The advantage of fuzzy logic for this case would be using fewer, simpler rules to handle all the cases reasonably. With fuzzy logic you could write three to six rules for size, motion, etc., and sum them. The rules follow more closely what a truck driver standing behind a truck does: instead of examining every item and plugging in every rule he or she knows about objects and decides that something is either 'a problem' or 'can be ignored' or 'keep an eye on it' as the driver is guided backwards. This allows them to ignore cats, birds, open lunch boxes, paper bags, ignore pedestrians who obviously see the truck and are moving at a safe distance from it without having to make a special rule for each of them. Things like drunks stumbling under the wheels, broken glass, other trucks or unaware pedestrians receive more attention because they fit into the class of 'problem items'.

In one sense all measured inputs have some fuzziness even in classical control. The mechanisms we have available for sensing, feedback and control all have limits to precision. As a result no rules will be 'absolutely accurate', since the error terms, physical limitations (e.g., phase-shift and attenuation) and input noise will normally be a noticeable part of the input. Also filtering inevitably reduces the legitimate input, attenuating the original signal. At that point, fuzzy logic simply recognises what we have been doing all along and 'hard' rules are ignored: mapping approximations onto the response which makes the useful outcome more likely.

3.5.6 Combination of fuzzy and conventional PID controllers

I understand that a fuzzy controller can be particularly good as an operator replacement. In this case can the rules be formulated by observing an operator's action?

Yes. We will consider a rules formulation problem in Section 4.5 in greater detail.

However, an operator in process control systems usually controls different technical devices including PID-controllers. Can a fuzzy controller manage this role?

Sure it can. In this case a fuzzy controller is placed on a higher, more intelligent level. It produces command signals for conventional controllers. Another possible task for a fuzzy controller in this structure is a future prediction. The famous Japanese subway and helicopter control systems are based on these principles.

Example 3.5 *Hybrid guidance control for a self-piloted aircraft [Bour96]*

A modern aircraft is well equipped with conventional control techniques and, in particular, various PID controllers which demonstrate a good performance and successfully solve different guidance problems. Guidance control in a modern aircraft is performed by the PID controllers producing the control signals which are applied to ailerons and elevators. The necessary reference inputs for a PID controller are usually supplied by the aircraft crew based on different data, first of all the current position provided through the global positioning system (GPS).

Some piloted aircraft classes are to be replaced with autonomous vehicles which are cheaper in operation and have some other advantages. The hybrid guidance control system, incorporating conventional PID controllers and a fuzzy controller, is proposed for this aircraft. A fuzzy controller takes the place of a pilot (an operator) in developing reference signals for a PID controller.

The navigation of the self-piloted vehicle is organised by the onboard GPS receiver tied to a PC-based flight director. Flight-planning software generates a list of consecutive points necessary to track the mission-determined flight path. Onboard autopilots keep the aircraft stable, while the flight director (guidance system) interprets the point positions to determine the course, speed, climb rate and turns of the aircraft. The objective of the guidance system is to bring the aircraft to the next operational point at a specified altitude and to stabilise the vehicle to allow for the operation of the onboard photographic and/or measurement equipment. The aircraft is assumed to be guided to an initial position during the take-off stage before the guidance system takes over a control.

The designed control structure is shown schematically in Fig. 3.19. The altitude of the aircraft is controlled by a low-level conventional PID feedback controller through aerodynamic ailerons and an elevator with mechanical limits of their deflection angles. A speed of the aircraft is controlled by the throttle setting. The guidance fuzzy controller has to provide reference signals for the PID controller, which are required roll and pitch angles (ϕ_{ref} and θ_{ref} respectively) for a levelled flight, and also to produce the throttle setting command T_{th} if a change of altitude is required.

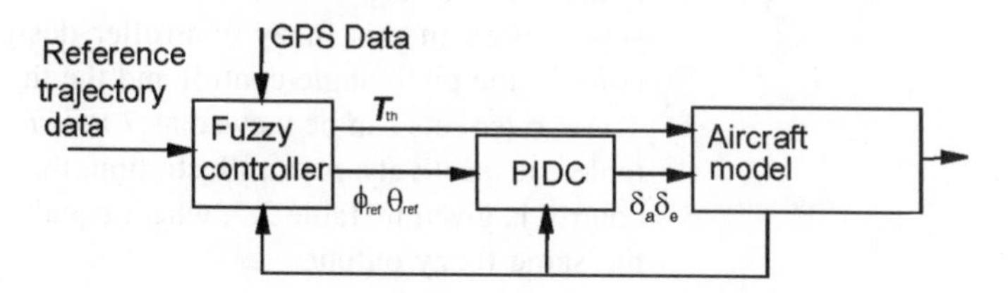

Fig.3.19 *Structure of the combined control system*

An operation of the fuzzy controller developed is illustrated in Fig. 3.20. Coordinates of the aircraft current position and of the next operational point are used to estimate an offset angle, $\dot{\delta}$, between the direction to the operational point and the current velocity vector, $\bar{v}$, and the rate of change of the offset angle, $\dot{\delta}$, as well as an altitude difference between the current position of the aircraft and the operational altitude, h, and the rate of change of the altitude difference $\dot{h}$, These estimates become the input signals for the fuzzy controller and are subject to fuzzification.

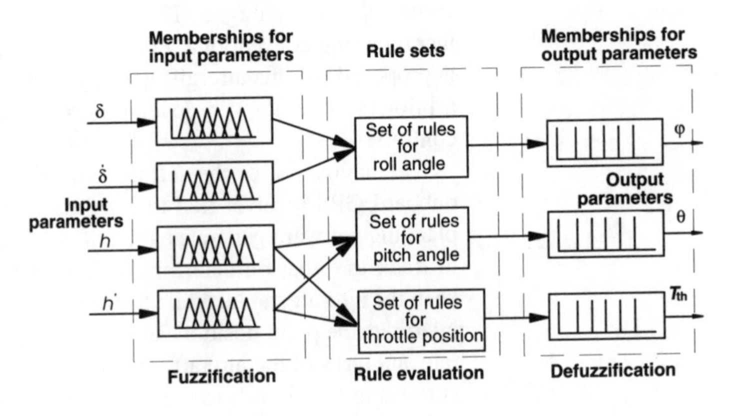

Fig. 3.20 *Operational block diagram of the fuzzy control system*

The whole control structure consists of three fuzzy controllers operating independently with each of them having two input and one output signals. This input–output mapping provides a simple two-dimensional structure of the linguistic rules sets.

The fuzzy controller output mainly depends on a definition of the membership functions and the rules. The control variables δ, $\dot{\delta}$ and h domains are divided into seven linguistic values with the relative memberships, and the control variable $\dot{h}$ into five, respectively. The rule definition is subjective and based on the expert's knowledge and experience. For a system with two control variables and seven membership functions in each range it may lead to a 7×7 decision table. The total of three rules sets is used in this fuzzy controller design: for the roll angle control, the pitch angle control and the throttle position control. These rules sets can be viewed as 7×7, 7×5 and 7×5 decision tables, respectively. As an illustration, the rules set for roll angle control is given in Table 3.7, where equally shaded cells produce the same fuzzy output.

Table 3.7

$\dot{\delta}$ \ δ	NB	NM	NS	SM	PS	PM	PB
NB	LB	LB	LB	LB	LB	LB	LL
NM	LB	LL	LL	LL	LL	LL	LM
NS	LL	LM	LM	LM	LM	LM	LLIT
SM	LM	LLIT	LLIT	LLIT	LLIT	LLIT	LMN
PS	LLIT	LMN	LMN	LMN	LMN	LMN	LLN
PM	LMN	LLN	LLN	LLN	LLN	LLN	LBN
PB	LLN	LBN	LBN	LBN	LBN	LBN	LBN

In this table linguistic labels for roll angle are denoted: LB, big; LL, large, LM medium; LLIT, little; the character N denotes negative.

Simplicity and low hardware implementation cost determine a choice of the membership functions of a singleton type for the output parameters (roll angle, pitch angle and throttle position). The linguistic variables of the fuzzy outputs that are evaluated by cycling through the rule sets are projected onto output sets of the memberships.

The defuzzification process takes place after the generation of the fuzzy control signals is completed using the inference mechanism. As more than one fuzzy output variable can be assigned a non zero degree, the contribution of each variable into the physical output should be taken into account. The defuzzification method was based on calculating the centre of gravity of all fuzzy outputs for each system physical output.

3.6 Stability and performance problems for a fuzzy control system

3.6.1 Stability and performance evaluation by observing the response

We understand now how a fuzzy controller operates. The next problem is how to decide if it functions well?

A control engineer often judges the quality of a control system by studying the response curve of the system. The curve reveals the dynamics of the system, responding to changes in its inputs.

So to make a decision, do we need to consider all possible inputs?

Of course not. Usually a small set of standard signals are studied (Fig. 3.21). A unit step input imitating a change of a set-point (Fig. 3.21a), a unit ramp input (Fig. 3.21b) simulating smooth change of the operating conditions, and rather seldom, a parabolic signal (Fig. 3.21c) when one needs to simulate a variable speed of changing.

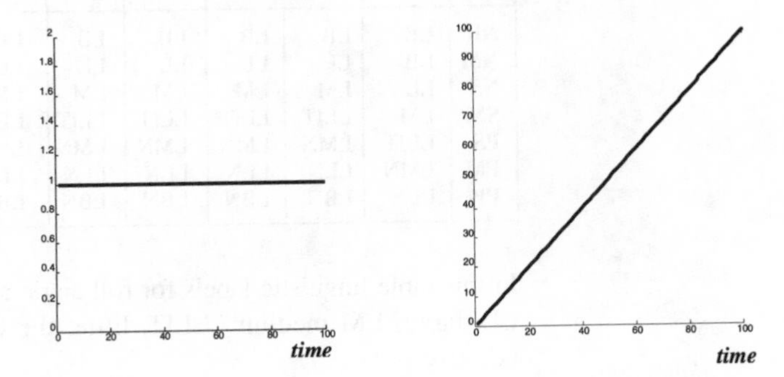

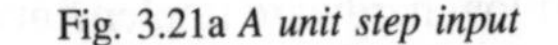
Fig. 3.21a *A unit step input*

Fig. 3.21b *A unit ramp input*

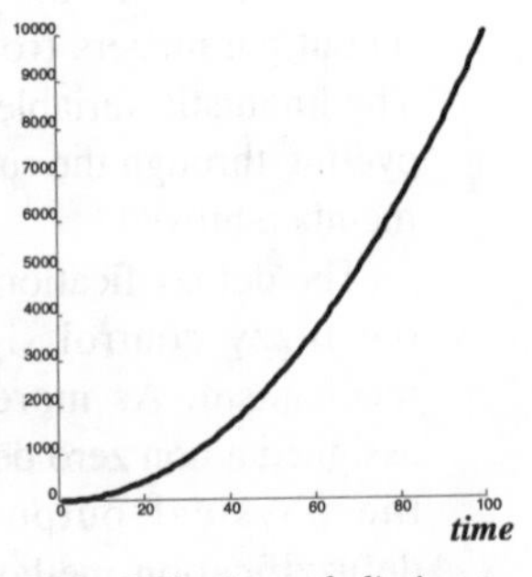

Fig. 3.21c *A parabolic input*

If a system responds well to these signals, a control practitioner usually judges it as a good system.

What does 'responds well' mean?

First of all, the system should prove to be stable. Stability is usually the most important issue to examine in determining the quality of the response (curve) of a system. It is concerned with the good behaviour of an output of the control system in response to either external inputs or internal excitations. Stability is a very complex theoretical problem, indeed, so complex that there are even different definitions of stability. Let us take the simplest one. The system is stable if its response for every bounded input is bounded. Practically, it means that for any finite change in its input, the system does not produce infinite output change. Or considering a unit step input the system response should be limited in magnitude. Figure 3.22a demonstrates some examples of a unit step response for a stable system, and Fig. 3.22b shows an unstable system.

I think it could be unreliable to analyse the system stability just by observing the response curve. Are there some other methods?

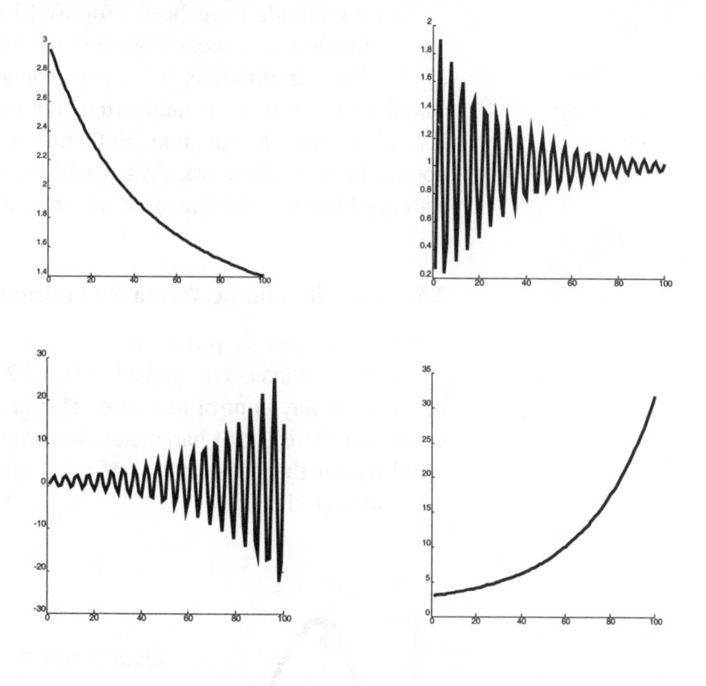

Fig. 3.22a *Examples of a unit step response for a stable system*

Fig. 3.22b. *Examples of a unit step response for an unstable system*

Yes, there are. In control theory there are some methods, among them:

- Nyquist's stability criteria;
- Routh's stability criteria;
- describing-function approach;
- phase-plane method.

Unfortunately all of these approaches cannot be used for stability analysis of fuzzy control systems.

Why not?

All of them require a special mathematical model of a control system prior to stability analysis. Nyquist and Routh–Hurwitz's criteria require a Laplace transform or z transform of a mathematical control model. The model is usually required to be linear due to the characteristics of the Laplace transform and z transform. Furthermore, these two criteria often fail if the control system is time-variant. The describing-function approach for determining stability is only approximate. The phase-plane method usually applies only to first- and second-order systems.

Are there any criteria to determine the stability of a fuzzy control system?

Some methods have been proposed based on the Liapunov second method. The second method of Liapunov (also referred to as the direct method) is the most general for determining the stability of a nonlinear and/or time-variant system of any order. It is a hot research topic now. But a universal reliable method has not yet been put forward. So I would recommend evaluation the stability by observing the system response and its parameters.

3.6.2 Stability and performance indicators

What parameters do you mean?

These parameters are marked in Fig. 3.23. Their definitions can be found in any control textbook. They are overshoot, rise time, settle time. This set of parameters is used to evaluate not just the stability, but the performance of a system as well, and often is given in specifications.

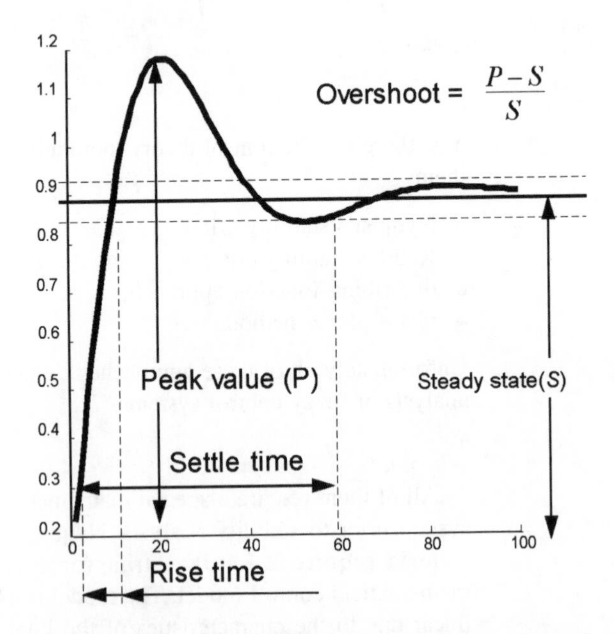

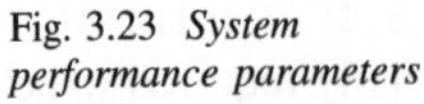
Fig. 3.23 *System performance parameters*

So to evaluate the performance, one has to determine the value of the indicators mentioned above.

Not always. One may apply other indicators. For example, steady state error. Or integral criteria.

How should I choose the indicator?

Generally speaking, the indicator choice depends on the problem domain.

Example 3.6 *A performance indicator for navigation control.*

In Example 3.5, the task was to direct an aircraft to the destination point. The final errors in positioning of the aircraft with respect to the target point and the attitude errors were taken as measures of performance. The orientation error was defined as the offset angle between the direction to the arrival point and the direction of the current velocity vector. Figure 3.24 gives an example of the fuzzy control results. In the plot of positioning errors, the distribution is elliptically spread, indicating a normal distribution of the coordinate errors.

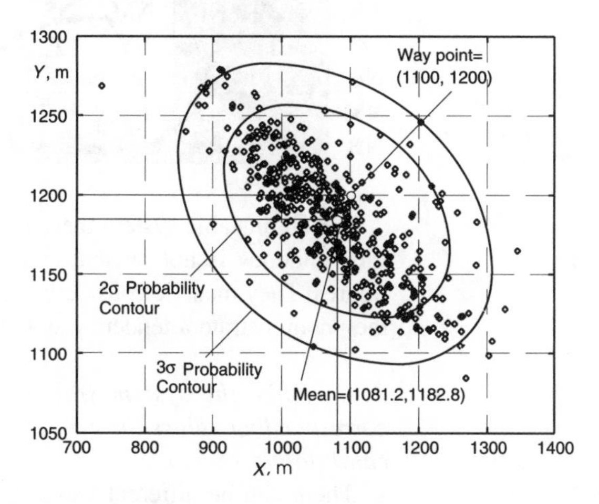

Fig. 3.24 *Arrival points distribution for fuzzy control*

And what about the stability? When is the stability not good?

Control systems have a high potential to be unstable if the following are observed in the response curve of the system:

- very fast rise time;
- high overshoot ratio;
- very long settle time.

To improve the system stability, one needs to decrease the overshoot ratio and the settle time.

How can I do that?

You can adjust either the rules, or the input and output scaling factors, or some other parameters of a fuzzy controller (See Chapter 5).

3.6.3 Stability evaluation by observing the trajectory

There is another method of predicting the stability of the system – observing the order in which the rules are fired. If we draw this order as a trajectory on the rules table for a good stable system, the error and change of error should be gradually decreased during the rules firing process. It means that we have to move along this trajectory from the edges of the table to its centre (Fig. 3.25) decreasing the group number from Group 4 towards Group 0.

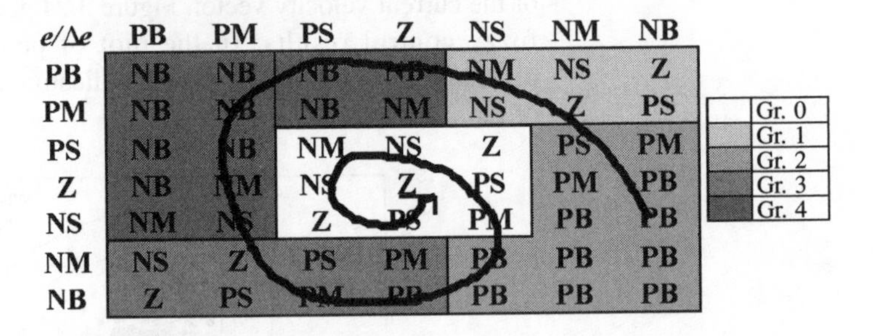

e/Δe	PB	PM	PS	Z	NS	NM	NB
PB	NB	NB	NB	NB	NM	NS	Z
PM	NB	NB	NB	NM	NS	Z	PS
PS	NB	NB	NM	NS	Z	PS	PM
Z	NB	NM	NS	Z	PS	PM	PB
NS	NM	NS	Z	PS	PM	PB	PB
NM	NS	Z	PS	PM	PB	PB	PB
NB	Z	PS	PM	PB	PB	PB	PB

Fig. 3.25 *A typical trajectory of rules firing for a stable system*

For an unstable system there is an inverse tendency or at least this tendency is not present (Fig. 3.26). Be careful with these figures. They must be considered as just examples. And of course, they demonstrate a tendency, not an exact order to follow.

So to make the system stable we need just to provide the corresponding values for the parameters mentioned above. How can I do that?

There can be different ways. The first and most obvious one is to try to avoid making mistakes in fuzzy controller design. The wrong choice of the rules and/or membership functions (to a lesser degree) may make the system unstable. The first thing which should be done is check the rules table. The widespread error for the beginner is the wrong sign for the table output or one of the inputs. It is easy to correct such a mistake. One needs just to replace ***NegBig*** with ***PosBig***, ***NegSmall*** with ***PosSmall***, ***PosBig*** with ***NegBig*** and so on. Another mistake is the wrong choice of the range for possible input or output values.

If one makes no design mistakes, will the system be stable?

It is not so simple. Even if one makes no wrong decisions in the simple structure controller design, the system may still be unstable.

e/Δe	PB	PM	PS	Z	NS	NM	NB
PB	NB	NB	NB	NB	NM	NS	Z
PM	NB	NB	NB	NM	NS	Z	PS
PS	NB	NB	NM	NS	Z	PS	PM
Z	NB	NM	NS	Z	PS	PM	PB
NS	NM	NS	Z	PS	PM	PB	PB
NM	NS	Z	PS	PM	PB	PB	PB
NB	Z	PS	PM	PB	PB	PB	PB

Gr. 0
Gr. 1
Gr. 2
Gr. 3
Gr. 4

Fig. 3. 26 *A typical trajectory of rules firing for a unstable system*

3.6.4 Hierarchical fuzzy controllers

What can be done in this case?

You may try another structure: try to apply a hierarchical (multilevel) fuzzy controller or a fuzzy controller with meta rules. In [Wang94] this approach is called a supervisory control. The idea is pretty simple. The fuzzy controller is designed to satisfy the performance specifications without a serious consideration of stability problems. As the trade-off between performance and stability is the most common problem in any controller design, this way one is able to simplify the controller design significantly as well as improve performance.

Hang on! Do we need a good performance when the system is or can become unstable?

No, we do not. This controller has to work and cope with ordinary situations. When the situation tends to become dangerous from the stability viewpoint another controller has to take control. This second controller is called a supervisory controller.

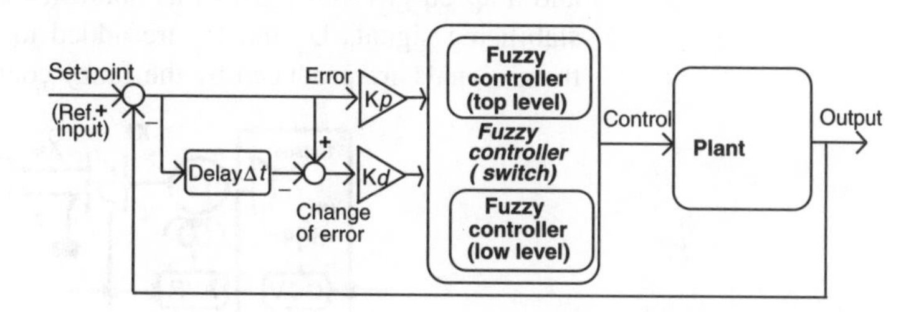

Fig. 3.27 *Multilevel structure of a fuzzy control system*

As the current theoretical methods of proving fuzzy control system stability are limited in their applications, the stable-in-theory top level controller is highly desirable. This demands the controller to be constructed as a simple unit with a small number of inputs and rules. The deviation error can possibly be the single input for this controller and it can be implemented with five to nine rules. This controller should be switched on when the deviation error is large, and off when it is small. We use rough terms to determine the switching rules as the switcher itself may be implemented as a fuzzy controller. Undoubtfully this controller should provide fast convergence of the output(s) to the desired values. The steady-state characteristics of this controller are not very important, because within a short time it will be switched off.

The low level controller(s) can have a complicated structure and achieves the desired steady-state and time response characteristics. In constructing these controllers, all the methods developed until now can be implemented. The intermediate level controller(s) should provide smooth handing over the control between the levels and eliminate the possibility of oscillation between them. The deviation error can be considered as the control input of the switch controller.

Example 3.7 *Fuzzy control system for power system stabilisation*

One of the main power system stability problems is the low-frequency oscillations of the interconnected electric power systems. These oscillations may be sustained for minutes and grow to cause a system separation, if no adequate damping is available. The model of one machine infinite-bus power system [Ishig92] is given in Fig. 3.28. Unsurprisingly, the design of a state-of-the-art power system stabiliser (PSS) for improving stability has received much attention. This stabiliser should demonstrate good performance under small as well as high disturbances under various operating conditions of the power system. The system is equipped with a fast-acting exciter (AVR) and a speed governor (GOV) as controllers. The supplementary stabilising signals U_a and U_g are added to each controller, and these signals are produced by the fuzzy controller.

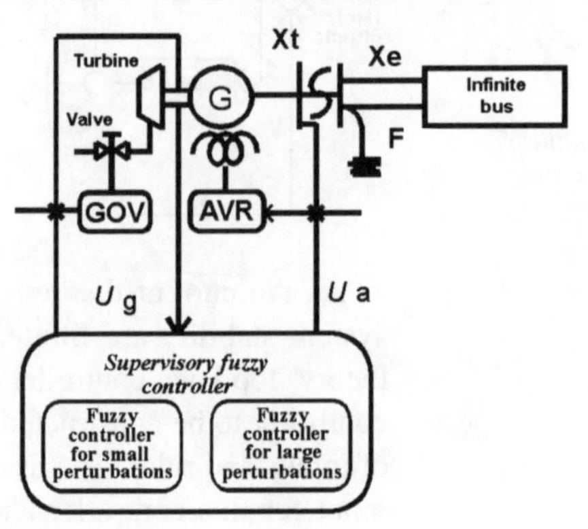

Fig. 3.28 *The model of a power system stabilised by a fuzzy controller [Ishig92]*

To solve this problem the structure of the fuzzy control system was proposed and tested [Shi95]. The hierarchical fuzzy control system includes a multilayer controller which incorporates two fuzzy controllers on a low level and one on the upper level. Higher level supervisory control is used for supervising an operation of the direct controllers. Typical tasks of the supervisory controller are monitoring the power system to determine the operating modes, assignment of a direct controller and an adaptation scheme. There are two fuzzy controllers at the lower level: one is responsible for operation under large perturbation, the other under small perturbation.

The supervisory controller monitors a value of the frequency speed deviation. When the power system is subject to the large perturbation, the speed deviation is high and vice versa. The lower level fuzzy controller for the large perturbation has 49 rules to cover the whole operating range, the lower level controller has just 24 rules. The dynamic responses of all fuzzy controllers are given in Fig. 3.29. Comparison analysis has proved that the hierarchical fuzzy controller provides the same performance as the 24-rule fuzzy controller under the small disturbance condition, however, it provides much better response than the 24-rule controller under the large disturbance condition.

Will this control system guarantee the stability for any system?

That is a good question. It is hard to answer in respect to any system. [Wang 94] has proved the stability of the hierarchical system satisfying some conditions. Other research has also proved the stability of a single fuzzy controller under some conditions.

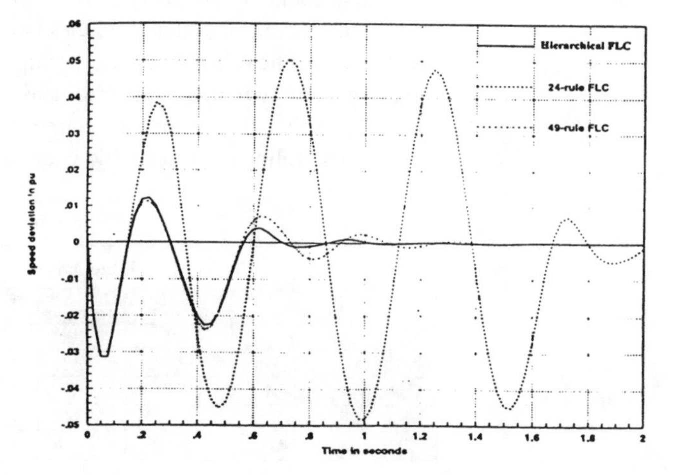

Fig. 3.29 *Speed deviation corresponding to a large disturbance of a three phase to ground fault*

What is the difference? What do we need this multilevel structure for?

Theoretically these conditions are different. Practically the main advantage is the design simplicity. It is very convenient to design a fuzzy controller based on just one criterion: either stability or performance, not both.

Example 3.8 *Servo controller for an optical disk drive [Yen92]*

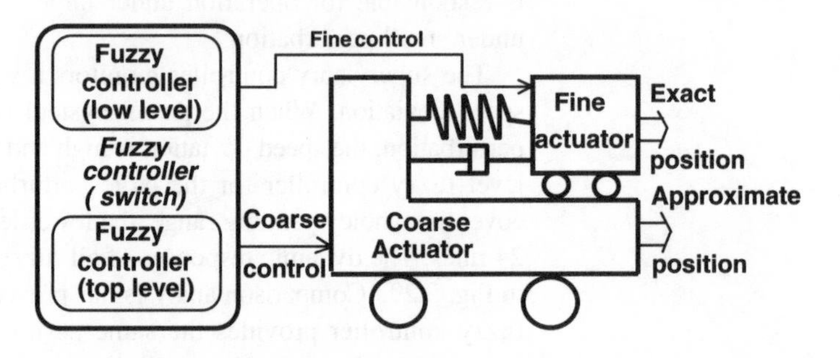

Fig. 3.30 *The model of a compound actuator with the coarse actuator and the fine actuator controlled by a fuzzy controller [Yen92].*

To control a disk drive, the compound actuator is widely used. Data in an optical disk drive is recorded on the disk surface as small areas of different reflectivity along concentric or spiral tracks. The data tracks are 2 mm apart. With a 5¼-inch disk, the actuator has to move the head across a 5cm stroke and position it with a 0.1μm error. The compound actuator consists of a fine actuator mounted on the back of a coarse actuator (see Fig. 3.30). The fine actuator has a fast response and a high resolution, while the coarse actuator can easily move over large distances.

The fuzzy control system includes two fuzzy controllers. Both are rather identical in respect to the inputs, output and the rules table. They use two inputs: the error and the change-of-error. Both inputs have five classes and the output (control signal) has nine classes. The following rules table is applied.

Table 3.8

e \ Δe	NB	NS	Z	PS	PB
NB	Z	PS	PM	PL	PG
NS	NS	Z	PS	PM	PL
Z	NM	NS	Z	PS	PM
PS	NL	NM	NS	Z	PS
PB	NG	NL	NM	NS	Z

The difference between fuzzy controllers is in the choice of the input and output scaling factors. The hierarchical structure is also applied here. To control the compound actuator the following rules are applied:

- if the fine actuator is capable of handling the head movement, it is applied,
- if the distance exceeds the fine actuator stroke, then the fine actuator is maintained at the stretched position and the coarse actuator is applied.

The two rules regulate the handover process between the two controllers.

HOW TO MAKE IT WORK OR DESIGN AND IMPLEMENTATION OF FUZZY CONTROLLERS

4 FUZZY CONTROLLER PARAMETER CHOICE

4.1 Practical examples

4.1.1 Fuzzy autopilot for a small marine vessel

Let us now reconsider our initial design attempt in a more formalised way. We want to design an autopilot fuzzy controller to control a movement of a small marine vessel.

Automation of a ship-steering process was initiated during the 1920s. Most modern autopilots are based on the PID controller and have fixed parameters that meet specified conditions [Polk93]. The aim is to keep a vessel moving along the predetermined trackline with as small a deviation as possible. A PID controller works well with a linear plant. In practice, maritime vessels are nonlinear systems. Any changes in the speed, water depth or mass may cause a change in their dynamic characteristics. In addition, the weather conditions will alter the effects of disturbances caused by winds, waves and currents.

Typical PID autopilots have settings to adjust course and rudder deadbands to compensate for vessel or environmental changes. Despite this possibility, the resulting performance is often far from optimal, causing excessive fuel consumption and rudder wear. These effects are particularly obvious in small vessels, where the sensitivity to disturbances and controller settings is far greater than with large ships. Fuzzy controllers are thought to be robust, enabling them to cope with nonlinear changes arising in ship dynamics and sea conditions.

We will design a fuzzy autopilot for a vessel which proceeds at a forward speed randomly distributed around 3.5 m/s and is to follow straight line tracks with as small a deviation as possible. A trackline may be up to 10 km in length. Any number of tracklines may be defined, and they may be parallel or directed at some arbitrary angle to the previous trackline. Parallel tracklines must be separated by at least 15 m.

The vessel is equipped with a position fixing system which will establish its position in x and y coordinates with normally distributed random errors (standard deviation of 1 m). The heading of the boat is measured with a normally distributed random error (standard deviation of 1 degree).

Table 4.1

Sea state	Disturbance		
	Boat heading w.r.t. current heading (degrees)	Position w.r.t. current position (m)	3 dB bandwidth of disturbance (Hz)
Dead calm	0	0	0.000
Slight	5	2	0.200
Moderate	10	5	0.125
High	20	10	0.100

The guidance requirements are to be accomplished in the presence of environmental disturbances of wind, waves and varying ocean currents. For the purpose of this problem, the influence of wind and waves may be treated as bandlimited random noise affecting the vessel heading and position. Disturbances as high as 10 m and 20 degrees may be assumed. The controller should be exercised over a range of environmental disturbances as defined in Table 4.1.

Ocean currents are constant within a given trackline and may be up to 1.5 m/s and may approach the vessel from any randomly chosen direction.

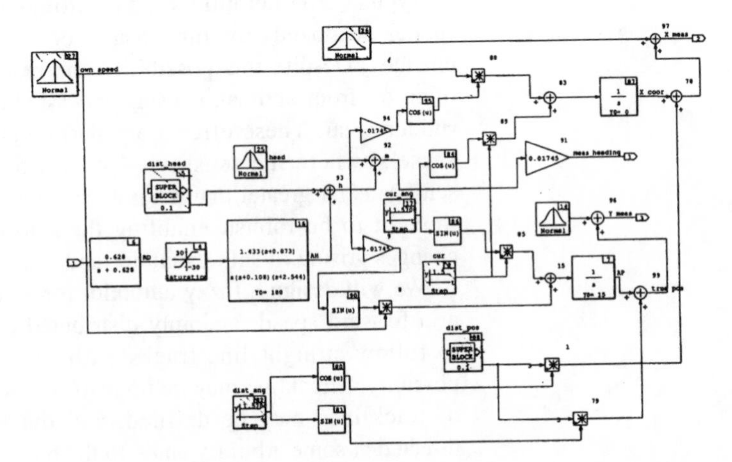

Fig. 4.1 *The MATRIXx™ simulation model for the vessel movement*

To give you more information the simulation model of the vessel movement process in MATRIXx is provided (Fig. 4.1). Because the heading angle and both coordinates can be measured, this model has three outputs: the angle (Output 1), and the vessel position coordinates, x (Output 2) and y (Output 3). This block has just one input: the control input, the rudder angle. In this model, blocks 4, 6, 16 perform the transformation of the rudder angle into the heading angle. Actually they simulate a 'pure' (without any disturbance and errors) reaction of the vessel heading on the control signal.

A few blocks simulate disturbance which acts on the heading (block 5) and position (blocks 98, 80, 81, 82). The Disturbance_for_position signal is produced by the block DIST_POS and is presented by the filtered random signal, the characteristics of which are chosen to satisfy the specifications (see Table 4.1). This block has three outputs to simulate slight, moderate and high disturbance. To imagine the influence of this disturbance you may have a look at Fig. 4.2 which represents one coordinate position disturbance only. The structure of the Disturbance_for_heading block is similar (block DIST_HEAD). This disturbance affects the heading angle of the vessel. The current is modelled as having a constant value, but the angle between the trackline and the ocean current direction can be changed. All measured values include measurement errors. These errors are modelled as normally distributed random values and are added to the unmeasured values to get real results.

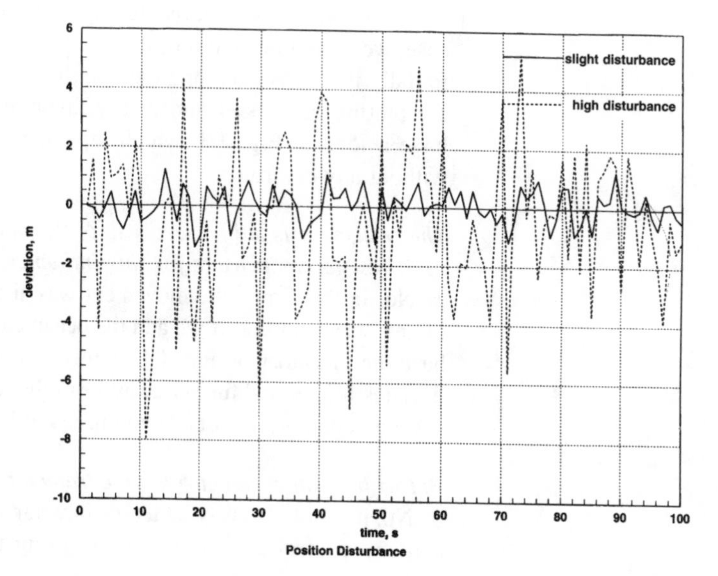

Fig. 4.2 *The simulation result for the Disturbance_for_position*

All these disturbances are used to simulate a real-life movement. The output of the vessel model (the heading angle) is affected by the Disturbance_for_heading (block 5) and the measurement error (block 25) to obtain the Heading_angle as the Output_1 for the block. Also this angle is used to get the movement projections on x and y axes after integration (blocks 87, 7). To simulate the movement, the random signal (normal distribution with the average 3.5 m/s and various standard deviations) that is modelling the actual speed of the boat is used (block 27). To simulate the real life movement, the signals modelling the Disturbance_for_position (blocks 98, 80, 81, 82), Ocean_current (blocks 2, 17, 84, 86) and measurement errors (blocks 14, 26) are added to get x (Output_2) and y (Output_3) coordinates of the current vessel position.

The values of the Ocean_current magnitude and angle and the Disturbance_for_position angle can be changed while modelling to allow for various cases. The initial conditions in the integrators may be changed to model different positions of the object before the movement starts. The initial condition in the transfer function block allows the model different initial positions of the heading in respect to the trackline direction before the movement starts. This model will be used as a plant model to design a fuzzy controller.

Do we need a controller? All the disturbances look pretty symmetrical. Will not the influence of any disturbance now be compensated for later on?

Suppose our vessel starts moving along the straight line. In this case, we do not need to turn it to keep the trackline. Suppose that initially it is directed along the trackline as well. In this case, we are putting our vessel under very favourable conditions. Let us have a look at Fig. 4.3, which represents the vessel trajectory without any controller.

The trajectory is shifted because of the ocean current effect. We need just to compensate for this influence.

No. In this figure the movement was simulated in the absence of any ocean current. If we add the ocean current of 1 m/s, we will have the trajectory on Fig. 4.4. In this case you see that our vessel deviates further and further away from the required trajectory. The system is definitely unstable and needs to be controlled.

But without an ocean current, the trajectory is not that bad, is it?

No, it is not. Anyway let us see how far we can improve it with a fuzzy controller, which we are going to design in the next sections.

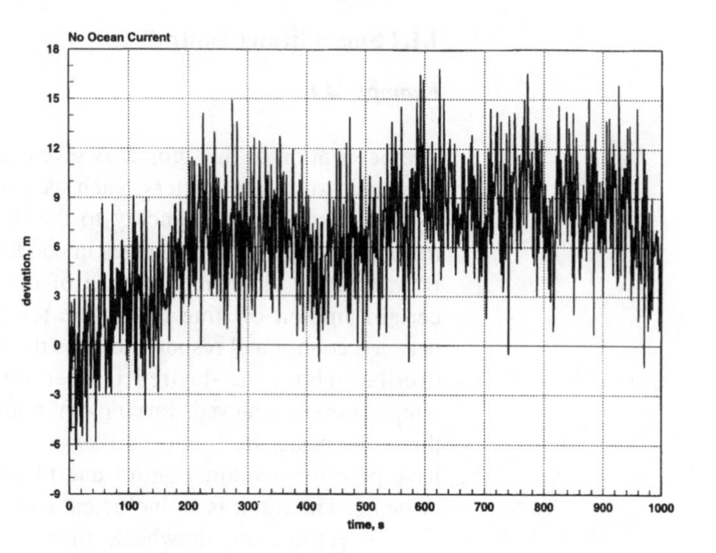

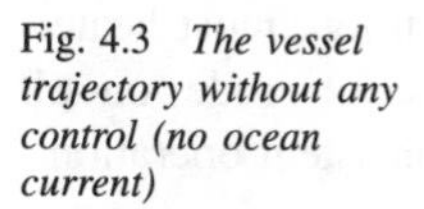
Fig. 4.3 *The vessel trajectory without any control (no ocean current)*

Why are we considering just a simulation? Real life examples are more interesting.

OK. Before we start improving let me present two real projects which have been developed and actually implemented by Adaptive Logic, Inc.

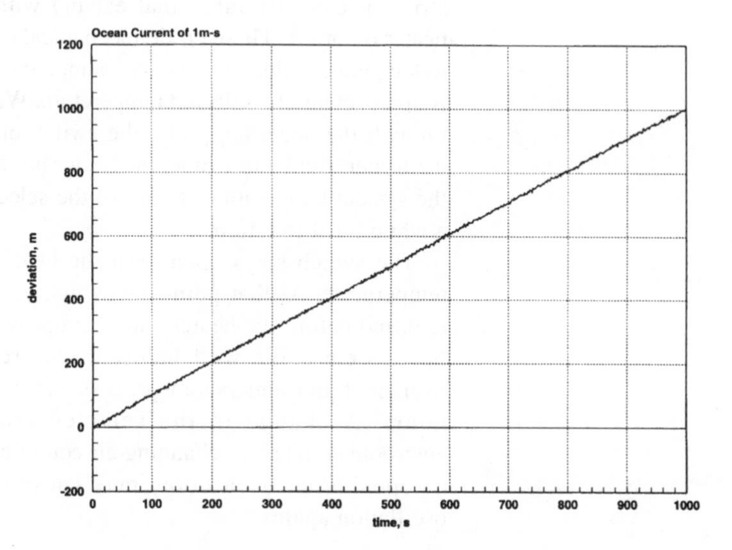

Fig. 4.4 *The vessel trajectory without any control (with the ocean current of 1m/s)*

4.1.2 Smart heater control

Example 4.1

The description of this project is given in [Adap96]. Household appliances that use heaters, such as ovens, rice cookers, and toasters, should come quickly to the desired temperature and maintain it regardless of changes in conditions such as load or air flow. Heating elements, because of thermal inertia, require a certain amount of time to change temperature. This latency between control and response causes the heater to overshoot and oscillate about the desired temperature. Once the selected temperature is achieved, varying environmental conditions often throw the heater back into oscillation. In addition, most heaters have poor temperature control due to the use of crude, on–off switches. The result is an inefficient and inconsistent operation.

To overcome this drawback, many industrial and consumer products as diverse as industrial chambers, ovens, rice cookers, toasters and irons use heating elements that could benefit from the advantages and feature enhancements of fuzzy logic control. A heater control has been designed that uses a fuzzy controller to sense the temperature, compare it to the user-selected temperature and control heater current. The inexpensive, dedicated fuzzy logic device is an adaptive controller that prevents thermal instability, providing consistent performance under all conditions.

We will describe both the fuzzy logic rules and design for the heater control. Most appliance heaters are controlled by bimetallic strips or capillary tubes that expand with heat and switch the heater on or off. These crude mechanical controllers merely react to temperature fluctuations and cannot anticipate when the heater is approaching the selected temperature. When the element passes through the operating point, the switch opens, cutting power to the heater. But by this time, the heater has enough energy to carry the system temperature far above the selected range and it takes a while for it to return.

The switch stays open until the heater cools to the correct temperature. At that point the switch closes, but some time is required before the heater can again provide sufficient heat and the system cools well below the correct temperature. The overshoot and undershoot process can continue for minutes or hours. A change in the selected temperature or in the environment (such as changing air conditions in a heating system or opening an oven door) may cause the heater to go into oscillation again.

Heaters have widely varying characteristics. They are specified in terms of their form including length, shape, thickness and material composition. Heating elements may add instability to a system because of their slow response time and thermal inertia. The heater specifications are based on the requirements of the end product. The end product also has a range of thermal characteristics that influence the behaviour of the heater. Many variables of heating systems make the design of a controller for different systems a difficult task. A control system using a fuzzy controller brings the temperature of the heater to the selected temperature quickly and keeps it there regardless of any changes in the load or environment. This results in a more stable and reliable operating temperature.

There are three external inputs monitored by the controller. The first comes from a thermistor to monitor the temperature. The second is the user-selected, desired temperature setting. Input three is, again, the measured thermistor value signal only delayed by a small amount of time. This last input enables the controller to know the direction and magnitude of the temperature change in addition to the absolute temperature. The controller samples the input data, processes it and outputs a pulse width modulated (PWM) output signal that switches a triac controlling the current through the heater.

The fuzzy controller design parameters (inputs, outputs, fuzzy variables and rules) are given in Fig. 4.5, followed by a brief description of the fuzzy controller operation. The main part of any fuzzy controller is implemented as a set of rules. It performs the control algorithm. By studying the rules, one can see the criteria for taking actions such as switching the heater on or off. These rules make decisions based on adjustable membership function definitions. The rules are easily modified to respond to different criteria. The following describes the rules' purposes in relation to the inputs, their associated fuzzy variables, and the action taken when a rule is fired.

Timer rules (Rules 1 and 2) are used to generate the timer for pulse switching the triac and to adjust the data processing rate. Rule one increments the ramp output if it is in the count membership function. When the ramp reaches the reset function, then the reset rule will be fired and return the ramp to zero to begin to increment through the count again. The rate of an increment is set to 12, but could be any non-zero value according to the requirements of the application. The increment and reset actions implement a timer that causes the ramp to sweep across an axis.

The heater output is used to define the centre value of membership functions On and Off. On and Off are, in turn, used by later rules to switch the triac on and off. Rules 3–19 consider both the current temperature and the previous value of temperature whenever the time has reached reset. The winning rule in this group will add to or subtract from the value of the heater.

The fuzzy variables classify the temperature as matching the selected value, or being too hot or cold. The membership functions used the user selected temperature (TSET) as a floating centre value to compare the selected temperature with the actual temperature. As the user varies the desired temperature value, the membership function centres move left or right. Other fuzzy variables use the current temperature value to define their centres. As the temperature varies, the functions shift right or left. The fuzzy variables compare the current temperature with its time-delayed value. The comparison is a calculation of the derivative value and sign of the temperature. By comparing both the current value and derivative of temperature with the desired temperature, one can calculate a temperature correction value based on both the absolute difference and the rate of change of temperature. The calculation allows for precise control and minimises the overshoot.

Rules 3–19 use one fuzzy variable from each set to adjust the value of the heater. The actions of the rules are designed to move the heater towards the desired temperature without causing it to overshoot. Table 4.2 summarises the various conditions and actions of the heater control rules. Note, if all of these rules are not fired (i.e., the blank boxes scenarios in Table 4.2) then the value of the heater is not changed. Some of the rules are described below. Values indicated are added from the heater based upon the winning condition of TEMP and DELAY. No action is taken where there are blanks.

Table 4.2

TEMP / DELAY	TVLOW	TMLOW	TSLOW	TON	TSHIGHT	MHIGH	TVHIGH
DLPOS		–15					
DSPOS		–8	–6	–1	–2		
DON	25	5	1	0	–1	–5	–25
DSNEG			2	1	6	8	
DLNEG						15	

Inputs
TEMP Measured temperature
TSet Desired temperature setting
Delay Delayed measured temperature

Outputs
Ramp Pulse width ramp
Heater PWM adjust value
Triac PWM output to triac

Fuzzy variables
Heater Triac Control
Ramp is Count
Ramp is Reset

Timing control
Ramp is On
Ramp is Off

Temperature detection
TEMP is TON
TEMP is TSHigh
TEMP is TMHigh
TEMP is TVHigh
TEMP is TVLow
TEMP is TMLow
TEMP is TSLow

Over-temperature detection
TEMP is OverT

Delayed temperature detection
Delay is DLNeg
Delay is DLPos
Delay is DSNeg
Delay is DSPos
Delay is DOn

Rules
Timing Ramp Generation
1. If Ramp is Count then Ramp + 12
2. If Ramp is Reset then Ramp = 0

Heater temperature control
3. **If** TEMP is TON and Delay is DOn and Ramp is Reset **then** Heater + 0.
4. **If** TEMP is TVLow and Delay is DOn and Ramp is Reset **then** Heater + 25.
5. **If** TEMP is TVHigh and Delay is DOn and Ramp is Reset **then** Heater –25.
6. **If** TEMP is TMLow and Delay is DLPos and Ramp is Reset **then** Heater –15.
7. **If** TEMP is TMLow and Delay is DSPos and Ramp is Reset **then** Heater –8.
8. **If** TEMP is TMHigh and delay is DLNeg and Ramp is Reset **then** Heater + 15.
9. **If** TEMP is TMHigh and delay is DSNeg and Ramp is Reset **then** Heater + 8.
10. **If** TEMP is TMLow and Delay is DOn and Ramp is Reset **then** Heater + 5.
11. **If** TEMP is TMHigh and Delay is DOn and Ramp is Reset **then** Heater –5.
12. **If** TEMP is TSLow and Delay is DSNeg and Ramp is Reset **then** Heater + 2.
13. **If** TEMP is TSLow and Delay is DSPos and Ramp is Reset **then** Heater –6.
14. **If** TEMP is TSLow and Delay is DOn and Ramp is Reset **then** Heater + 1.
15. **If** TEMP is TSHigh and delay is DOn and Ramp is Reset **then** Heater –1.
16. **If** TEMP is TSHigh and Delay is DSNeg and Ramp is Reset **then** Heater + 6.
17. **If** TEMP is TSHigh and Delay is DSPos and Ramp is Reset **then** Heater –2.
18. **If** TEMP is TON and Delay is DSNeg and Ramp is Reset **then** Heater + 1.
19. **If** TEMP is TON and Delay is DSPos and Ramp is Reset **then** Heater –1.

Over-temperature turn-off
20. **If** TEMP is OverT **then** Heater –1.

Heater triac control
21. **If** Ramp is Off **then** Triac = 0.
22. **If** TEMP is OverT **then** Triac = 0.
23. **If** Ramp is On **then** Triac = 255.

Fig. 4.5 *Fuzzy controller for a smart heater*

Rule 3, for example, is true if the temperature and its previous value are in the selected range, and leaves the heater unchanged. Rules 4 and 5 detect the temperature to be far from the selected value and not changing. These rules make large increases or decreases of the heat. Rules 6 and 7 both consider a medium-low value for the temperature (TMLOW), but they consider different derivatives and have different action values. Rule 6 considers a large positive delayed value (DLPOS) while Rule 7 considers a small one (DSPOS). At first glance, the correction actions for the rules may seem to be counterintuitive. In each rule the temperature is too low, yet the action is to reduce the value of the heater which decreases temperature when the heat is stable. The key is that in each rule the derivative is positive indicating that the temperature of the heater is increasing. The reduction in the value of heat prevents overshoot while still allowing the temperature to reach the set point.

The corrective action from Rule 6 is larger than that for Rule 7 because in the former rule the derivative value of the air temperature was larger indicating a faster increase in temperature. The larger previous value subtracts more from the heater (–15) than does the smaller (–8). The same reasoning holds in the remaining rules in the set. As the heater moves closer to the desired point, other rules will act to gradually hold it there.

Rules 20 and 22 guard against the heating unit from getting too hot. Rule 20 decreases the heater value in small increments when the temperature has gone above the safety level. During this process, at the same time, Rule 22 prevents the PWM control of the triac and keeps it off until the temperature decreases to below the safety value. Note that Rule 23 (which turns on the pulse) will never be enacted while the temperature is greater than the safety value. Even if the fuzzy variables **Ramp is ON** and **TEMP is OverT** should evaluate to the same value, the precedence goes to the first rule (Rule 22) for determining the action. If the temperature is above the safety value, **TEMP is OverT** will always be evaluated as a maximum value.

Rules 21–23 control the PWM output, triac, turning it on and off. The decision is based upon the comparison of the ramp feedback value to the heater value. The heater value serves as a centre for membership functions ON and OFF. As the ramp moves across count and back to zero, it also moves between the functions ON and OFF. Depending on the value of the heater, the position of the centre value for the two functions will move to the left or right. That will vary the amount of time during a given sweep that the heater is on or off. By moving the heater, one changes the width of the pulse controlling the triac and the heater temperature.

4.1.3 Active noise control

Another example is provided with the real application results.

Example 4.2
Perhaps, there is no more popular problem in telecommunication engineering than one of noise cancellation. One of the methods explored which is called an active noise control (ANC) is producing an anti-noise signal. Adaptive Logic, Inc. has developed a solution [Adap96] based on fuzzy controller design and realisation. The solution achieves audio noise reduction of over –20 dB. By adjusting phase and gain values from the measured noise level, the controller reduces significantly the

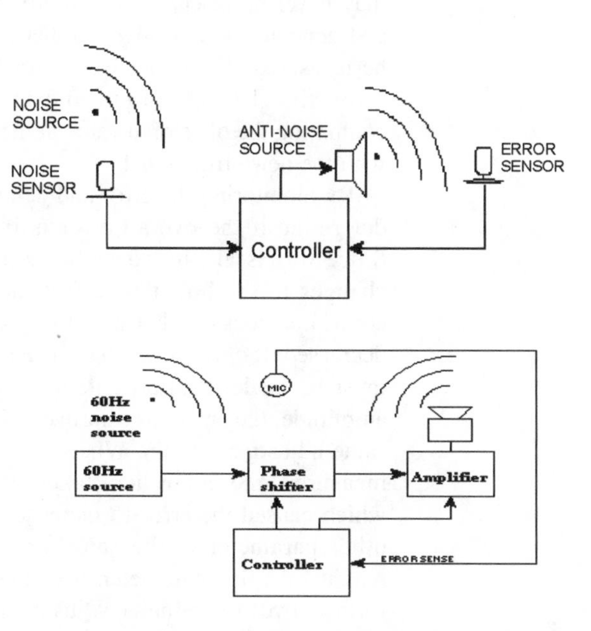

Fig. 4.6 *A typical ANC system*

Fig. 4.7 *Waveform synthesis for an ANC*

dynamic error measured at the cancellation point. The idea of an ANC is to produce an anti-noise signal coinciding with the noise in phase and magnitude to cancel it out. Figure 4.6 shows a typical ANC system that contains a noise source, noise sensor, error sensor and the control or anti-noise source. The system can be modelled as a closed loop system, where the error is continuously being minimised. Though there are various methods for producing the anti-noise, they all rely on the principle that a signal, when added to the inverse of itself, will produce a null. The problem is that in a time variable world, producing the exact inverse signal can be very difficult. Additionally, acoustic reflections can add to the original signal producing even more complexity in the system. The exact specifications for the anti-noise signal might be known for some very periodic systems. In these cases, only the phase and amplitude need to be adjusted. An example of such a system might be a large transformer or motor, which being driven at 60 Hz, produces 60 Hz noise and harmonics. For this system, the exact frequency is known so it can be regenerated, phase shifted and amplitude adjusted to make the anti-noise. Figure 4.7 shows a waveform synthesis system.

For simplicity in this example, the error signal is the RMS value of the noise sensor signal. Response time is critical, because if set too slow, transients will not be seen, and if too fast, a null may never be reached. The controller monitors the error signal, and generates a derivative, so that the direction of the error can be measured. The derivative is simply produced by copying the error signal to an output and then on the following cycle comparing the old stored value at the output (internally fed back) with the new error signal.

By monitoring the error and delayed error, the controller can determine if the overall error is improving or getting worse. Since this is all the controller knows, it must try parameter changes to see how they affect the error. In other words, the controller does not know if the phase should be increased or decreased. It only knows that since an error exists, an adjustment must be made. If after making an adjustment to the phase or amplitude, the error has improved, then the controller makes the same adjustment again. When an increase in the error signal is measured, the controller 'takes back' the last change made, which caused the error to increase. It then begins to adjust the other parameter in the same way before again returning to readjust the first parameter. It will continually switch back-and-forth between parameter adjustments tuning them for optimal noise cancellation. By making the adjustment larger when the error is large, and smaller when the error is small the time needed to reach a null can be shortened. Because the dynamics of this system are slow, audio frequencies, we must be sure that after making a change, we wait for the system to respond before the change to error is read.

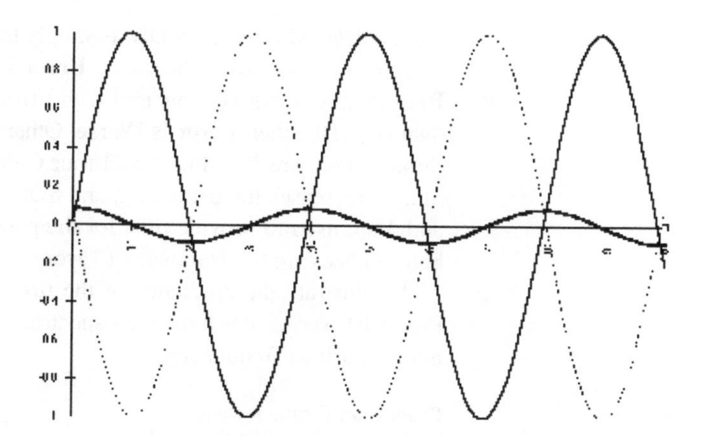

Fig. 4.8 *Error caused by a small phase shift*

If the remaining harmonics are high enough to cause concern, they too might need to be cancelled in a similar manner, but with a separate phase and amplitude circuit. For most systems of this type, only the primary frequency and first harmonic need to be cancelled.

The phase accuracy of the anti-noise causes the most change and therefore error in this type of system. Figure 4.8 illustrates a single wave being cancelled by its inverse, but with 5 degrees of error. It can be seen that with only this small degree of error, a significant signal still remains. This problem is only worsened with more complex signals, therefore care must be taken in the phase shifter design. Note that delay in a given system remains fairly constant over time. This means that the phase shifting circuit need not vary more than the delay would change over a long term, typically less than 45 degrees. So the resolution of the controller can be effectively increased by concentrating it to the narrower region.

The rule set for this example developed by Adaptive Logic, Inc. is listed in Fig. 4.9. This example uses one analog input, Error, and two analog outputs, Phase and Gain. Two other outputs, State and ErrorDly are used for internal variable storage. The State variable forms a state machine that controls the flow of the system. As discussed earlier, ErrorDly, is used to determine the direction of Error. In other words it tells the controller if Error is getting better or worse.

The rules are divided into four sections – one section for each of the two outputs and two variable registers. For each sample only one rule in each section can win. That rule then adjusts the output. If no rules are fired, the output does not change.

The delayed error, ErrorDly, is simply the Error delayed by one sample time. To solve the terms **Error is Worse** and **Error is BetterSame**, Error is compared with ErrorDly. If Error is greater than ErrorDly, then **Error is Worse**. Otherwise, it is **Better** or the **Same**. There are four Phase and four Gain states. In either case, State 2 increases the parameter, and State 3 decreases it. States 1 and 4 are used to provide time for the parameter change to settle before checking the response of Error.

To illustrate the efficiency of the fuzzy control methodology, Fig. 4.10 shows the frequency spectrum of a noise source with many spurious frequencies.

Phase control rules section

0. **If** State is Zero then Phase = 127.
1. **If** State is PHASESTATE2 and Error is High **then** Phase + 5.
2. **If** State is PHASESTATE3 and Error is High **then** Phase – 5.
3. **If** State is PHASESTATE2 **then** Phase + 1.
4. **If** State is PHASESTATE3 **then** Phase – 1.
5. **If** State is PHASESTATE1 and Error is Worse **then** Phase – 1.
6. **If** State is PHASESTATE4 and Error is Worse **then** Phase + 1.

Gain control rules section

7. **If** State is Zero then Gain = 127.
8. **If** State is GainState2 and Error is High **then** Gain + 5.
9. **If** State is GainState3 and Error is High **then** Gain – 5.
10. **If** State is GainState2 **then** Gain + 1.
11. **If** State is GainState3 **then** Gain – 1.
12. **If** State is GainState1 and Error is Worse **then** Gain – 1.
13. **If** State is GainState4 and Error is Worse **then** Gain + 1.

State control rules section

14. **If** State is Zero **then** State = PHASESTATE1.
15. **If** Error is Zero and State is NotZero **then** State = PHASESTATE1.
16. **If** State is PHASESTATE1 and Error is BetterSame **then** State = PHASESTATE2.
17. **If** State is PHASESTATE2 **then** State = PHASESTATE1.
18. **If** State is PHASESTATE1 and Error is Worse **then** State = PHASESTATE3.
19. **If** State is PHASESTATE3 **then** State = PHASESTATE4.
20. **If** State is PHASESTATE4 and Error is BetterSame **then** State = PHASESTATE3.
21. **If** State is PHASESTATE4 and Error is Worse **then** State = GainState1.
22. **If** State is GainState1 and Error is BetterSame **then** State = GainState2.
23. **If** State is GainState2 **then** State = GainState1.
24. **If** State is GainState1 and Error is Worse **then** State = GainState3.
25. **If** State is GainState3 **then** State = GainState4.
26. **If** State is GainState4 and Error is BetterSame **then** State = GainState3.
27. **If** State is GainState4 and Error is Worse **then** State = PHASESTATE1.

Error delayed rules section

28. **If** Anything **then** ErrorDly = Error.

Fig. 4.9 *Rule Set*

The frequencies, though centred around 500 Hz, extend down to 0 Hz and up to over 6 KHz. The noise level at the peak is about –2 dB. Figure 4.11 shows the cancelled spectrum for this noise source. With the active cancellation, the peak noise level is now at –25 dB and the energy contained in the other frequencies is dramatically reduced. With multinode cancellation other frequencies could be further reduced as well.

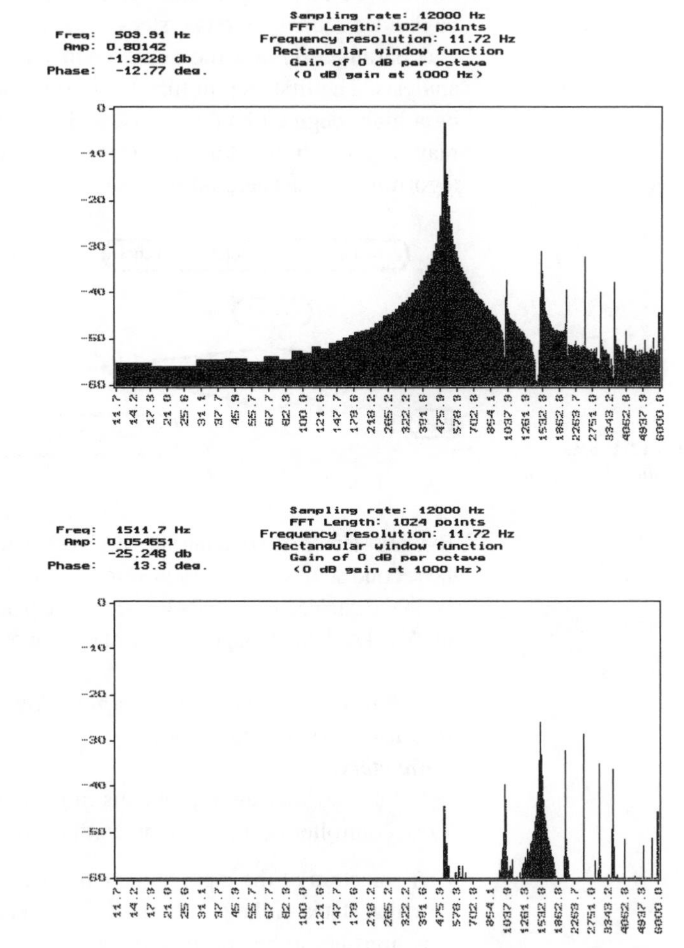

Fig. 4.10 *Complex noise source frequency spectrum*

Fig. 4.11 *Cancelled complex noise frequency spectrum*

4.2 Iterative nature of a fuzzy controller design process

A fuzzy controller design process contains the same steps as any other design process. One needs initially to choose the structure and parameters of a fuzzy controller, test a model or the controller itself and change the structure and/or parameters based on the test results. You may see that the actual design process consists of choosing the controller structure and some parameters (synthesis of the controller) and evaluation of their influence on the controller stability and performance (analysis of the controller). The processes of the analysis and synthesis are interrelated and interdependable on each other. The process can be divided roughly into two steps: an initial choice of the structure and parameters, and the subsequent adjustment based on the analysis. The first step in fuzzy controller design is characterised by a high degree of subjectivity, and as a result, the second step may require a high effort in order to be implemented. This simple algorithm is presented on Fig. 4.12.

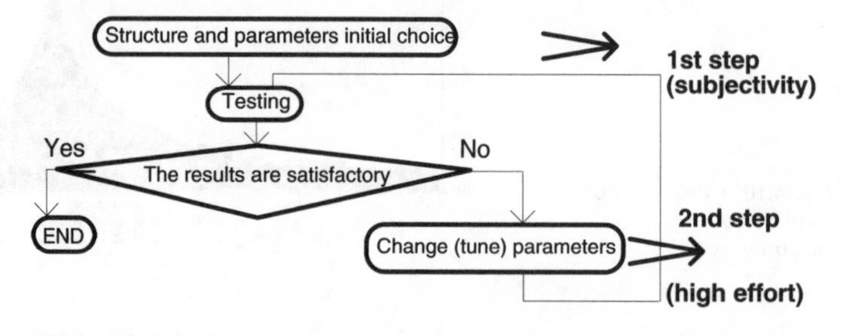

Fig. 4.12 *A fuzzy controller design process*

The first step of fuzzy controller design is considered in this chapter, the second step will be presented in Chapter 5. Obviously, the better the initial choice, the simpler the second step. One can hope to correct all mistakes later, though, a good start cannot be overrated.

This figure is really pretty simple. However, I have a lot of questions. For example, how do I choose parameters? Which parameters?

Let us consider these problems in greater detail. The classical fuzzy controller design scheme contains the following steps:

- A selection of the input and control variables – a definition of which states of the process shall be observed and which control actions are to be considered.
- A definition of the condition interface – fixing up the way in which observations of the process are expressed as fuzzy sets.
- A design of the rule base – a determination of which rules are to be applied under which conditions.
- A design of the computational unit – supplying algorithms to perform fuzzy computations. Those will generally lead to fuzzy outputs.
- A definition of the rules according to which fuzzy control statements can be transformed into crisp control actions.

The first step is usual in design of any control system. Inputs and outputs of the controller need to be chosen.

Should we apply the error and change of error as the inputs of the controller?

This is a universal approach for a feedback control in a case of a PD-like fuzzy controller. Sometimes another choice can produce better results. In our vessel movement control problem, both the deviation from the trajectory and the difference between the required heading angle and the actual one can be considered as an input. One or both of them can be used.

Hang on! You mentioned that both the position and angle of deviation can be chosen as the inputs. Which one should I choose?

I do not know a general answer. In Fig. 4.13 and Fig. 4.14, you see the movement trajectory for both angle and deviation taken as the inputs. However, the results presented on Fig 4.13 do not take into account the current and in Fig. 4.14 the current of 1 m/s is simulated. You see the results are pretty different. Because a fuzzy controller realises a nonlinear mapping, it is very hard to predict the performance of which controller and in which aspect would be better. I would advise you to try both of them. And if you design a fuzzy regulator fighting against high disturbances, try a parallel structure of both fuzzy controllers.

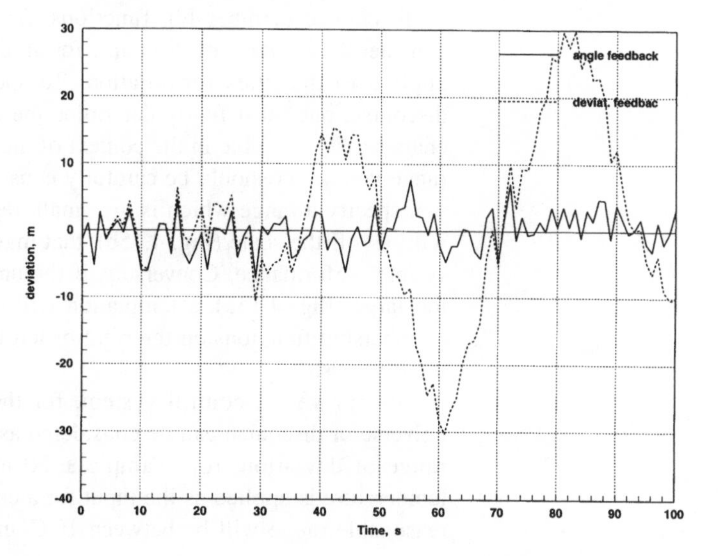

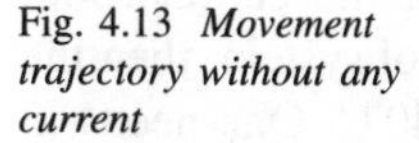
Fig. 4.13 *Movement trajectory without any current*

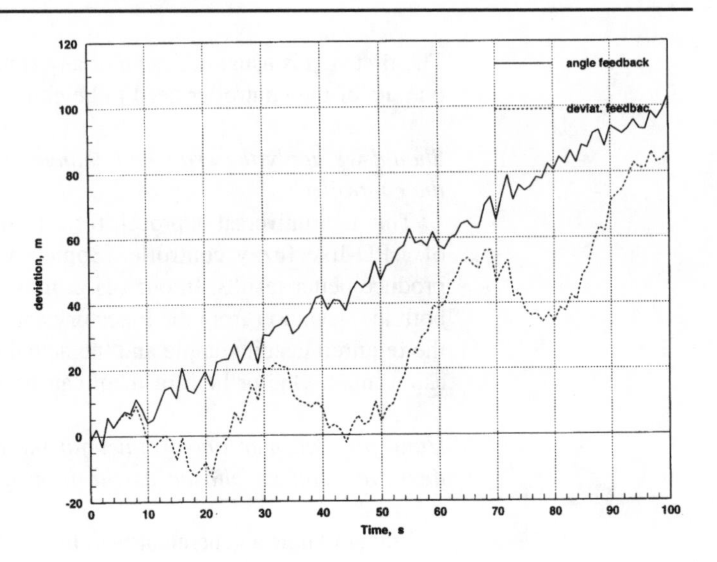

Fig. 4.14 *Movement trajectory with the current of 1 m/s*

4.3 Scaling factor choice

4.3.1 What is a scaling factor?

The first stage of the fuzzy controller operation is fuzzification and the last one is defuzzification. On both stages, the membership functions which describe different values of the linguistic variables (or labels) are applied.

What is a scaling factor?

To choose membership functions, first of all one needs to consider the universe of discourse for all the linguistic variables, applied to the rules formulation. To specify the universe of discourse, one must firstly determine the applicable range for a characteristic variable in the context of the system designed. The range you select should be carefully considered. For example, if you specify a range which is too small, regularly occurring data will be off the scale (Fig. 4.15b), that may impact on an overall system performance. Conversely, if the universe for the input is too large (Fig. 4.15a), a temptation will often be to have wide membership functions on the right or left to capture the extreme input values.

For our vessel control system, for the deviation input the universe of discourse can be considered as a possible reasonable range of deviation, for example ± 50 m. If the actual room temperature is applied as the input for a control system, then the reasonable range will be between 10°C and 40°C. One needs to

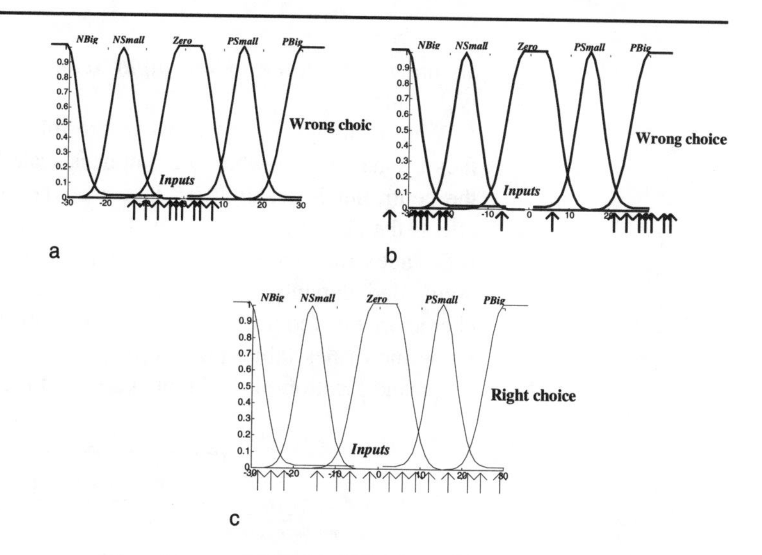

Fig. 4.15 *Choice of the scaling factors*

remember that a reasonable choice in one case can become bad in another. It is very hard to give general recommendations.

What can we do in this case?

Because of this situation it is usually desirable and often necessary to scale, or normalise, the universe of discourse of an input/output variable. Normalisation means applying the standard range of [–1,+1] for the universe of discourse both for the inputs and the outputs.

Doesn't it mean that our actual values will be too large or small?

It does if we do not modify the structure of the fuzzy controller. The structure of Fig. 3.10 should be replaced with one given in Fig. 4.16.

In the case of the normalised universe, an appropriate choice of specific operating areas requires scaling factors. An input scaling factor transforms a crisp input into a normalised input in order to keep its value within the universe. An output scaling factor provides a transformation of the defuzzified crisp output from the normalised universe of the controller output into an actual physical output.

What do these scaling factors affect?

The role of a right choice of input scaling factors is evidently shown by the fact that if your choice is bad, the actual operating area of the inputs will be transformed into a very narrow subset of the normalised universe or some values of the inputs will be saturated.

So the right choice of the input scaling factors is the most important thing.

Wait a minute. Let us consider the role of the output scaling factor. One can see when the output is scaled, the gain factor of the controller is scaled. The choice of the output scaling factor affects the closed loop gain, which as any control engineer knows, influences the system stability. The behaviour of the system controlled depends on the choice of the normalised transfer characteristics (control surface) of the controller. In the case of a predefined rules table, the control surface is determined by the shape and location of the input and output membership functions.

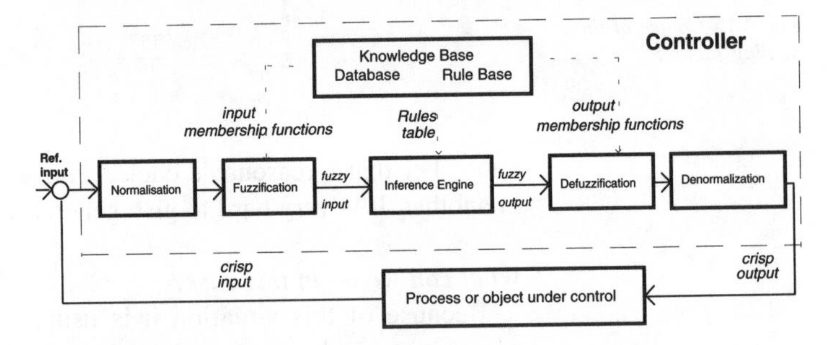

Fig. 4.16 *The structure of a fuzzy controller with the normalisation and denormalization blocks*

4.3.2 Where should the tuning start?

[Palm95] recommends the following priority list:

- The output denormalisation factor has the most influence on stability and oscillation tendency. Because of its strong impact on stability, this factor is assigned to the first priority in the design process.
- Input scaling factors have the most influence on basic sensitivity of the controller with respect to the optimal choice of the operating areas of the input signals. Therefore, input scaling factors are assigned the second priority.
- The shape and location of input and output membership functions may influence positively or negatively the behaviour of the controlled system in different areas of the state space provided that the operating areas of the signals are optimally chosen. Therefore, this aspect is the third priority.

According to these recommendations, I understand that we have to choose the output scaling factors and then the input scaling factors.

It is better to try to make both choices simultaneously. Actually all scaling factors influence the performance indicators.

Yes, like coefficients in PID controllers. Maybe our previous experience with them can help.

Absolutely! The obvious way is to apply the existing knowledge of classical PID design to a fuzzy controller design. This idea was proposed a rather long time ago [Tang87].

To do it we need to understand the similarity between a fuzzy controller and a PID controller. We have to analyse the relationship between the scaling factors of a fuzzy controller and the coefficients of a PID controller.

Exactly! Some research has been conducted in this area. The most clear recommendations are presented in [Yag94], which analysed a fuzzy PI-like controller (Fig. 3.14) in comparison with a classical PI controller. The analysis established the similarity between coefficients K_I and K_P of the PI controller and scaling factors and, proves that a fuzzy controller can be approximated by a virtual PI controller:

$$\Delta u(t) = K_I e(t) + K_P \Delta e(t)$$

with parameters:

$$K_I = K_I(K_{du} K_e) \text{ and } K_P = K_P(K_{du} K_d).$$

From here one can conclude that:

- increasing/decreasing the output scaling factor K_{du} causes increasing/decreasing both coefficients of the corresponding PI controller, and therefore, it determines the gain factor of the fuzzy controller;
- increasing/decreasing the input scaling factor K_e results in increasing/decreasing of the parameter K_I;
- increasing/decreasing the input scaling factor K_{de} causes increasing/decreasing of the parameter K_P.

How can it help in the fuzzy controller parameter choice?

If you do not have any experience in a classical PID controller design, it is no help. However, if one applies the knowledge of PID coefficients' influence on a controller performance, one will be able to use similar rules for tuning scaling factors of a fuzzy controller in the manner identical to a PI controller. The rules derived for tuning a PI-like fuzzy controller scaling factors are given in Table 4.3. A similar relationship can be derived for a PD-like and a PID-like fuzzy controller (see Fig. 4.17).

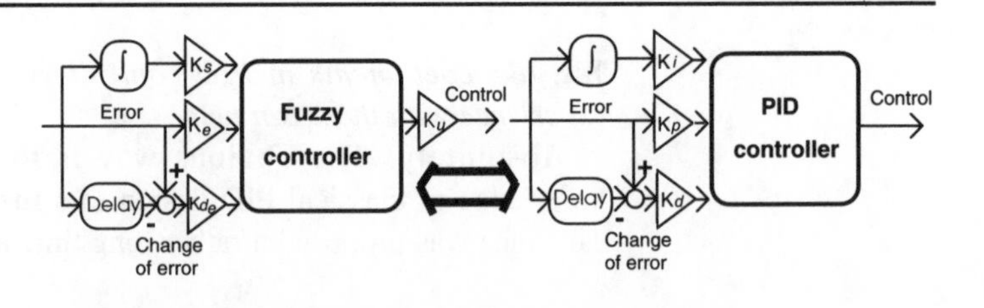

Fig. 4.17 *Similarity between scaling factors of a fuzzy controller and coefficients of a PID controller*

Table 4.3 Rules for tuning parameters of a PI-like fuzzy controller [Rez95-2]

	Tuning action for		
Error attribute	K_{de}	K_{du}	K_e
Steady divergence	Decrease	Decrease	Decrease
Overshoot / oscillation	Decrease	Decrease	Decrease
Speed of response	Increase	Increase	Decrease
Steady state error	Decrease	Decr/Inc	Increase

As far as I understand these recommendations, they give some advice of how to tune scaling factors or at least in what direction they should be shifted. However, they do not explain how to choose these parameters initially.

I do not agree. The relationship between a fuzzy controller and a PID controller given in [Yag94] has a global nature. One can use them as a basis for other recommendations. For example, one can see that the gain factor of a fuzzy controller could be evaluated approximately as a product of an output and a first input scaling factor. So initially one can choose the output scaling factor to provide the gain required by the controller specifications. Of course, design is a trial-and-error process. Because one understands how a change of the scaling factors influences the performance characteristics, one can try different scaling factors, changing their values in a corresponding direction, until the controller demonstrates the performance required by the specifications.

We discussed before that an output scaling factor is determined by the universe of discourse for the output, similarly for an input. Now you are talking about other reasons.

Sure, if you have a possible output range limited by specifications, physically or by any other way, you should not cross this boundary. One needs to remember that any controller, including a fuzzy one, produces an output signal, which is applied to the object under control. This object can impose some limitations upon the signal. In this case, the output scaling factor is fixed. Then the input scaling factor should be determined to provide the required gain for the controller.

Aren't the same limitations valid for the input?

If we use a feedback structure and apply an error signal as one of the controller inputs, the answer is generally no, because usually it is hard to draw up obvious limitations for the error signal. However, if we use another specified input, the limitations can be derived from the controller future behaviour consideration, and these limitations have to be applied to determine the scaling factors.

Sometimes the right choice of the scaling factor becomes the main problem of a design project. In this case, special attention is paid to the problem. The scaling factor can be determined, for example, at early stages of the fuzzy controller design and used later.

4.3.3 Application example

Example 4.3 A camera-tracking fuzzy controller [Lea92]

A control system for managing the traffic around a space station is under development at the Johnson Space Centre. The first phase of this development is a camera-tracking system that should control the camera gamble drive to keep any object in the field of view (FOV) of the camera.

The camera generates measurement results in terms of pixels where the field of view is represented with a 512×512 pixel map.

A fuzzy controller controls pan and tilt rates based on pixel object position.

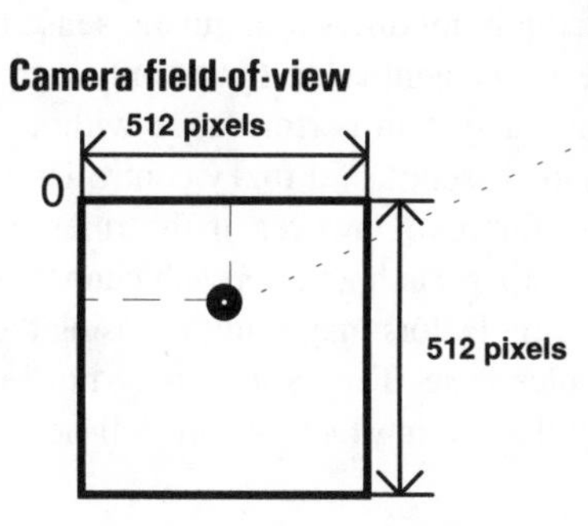

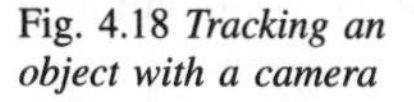

Fig. 4.18 *Tracking an object with a camera*

The object tracking means aligning the pointing axis of a camera along the object line of sight (LOS) vector. The monitoring camera is typically mounted on the pan and tilt gamble drives, which are capable of rotating the pointing axis within a certain range. The object LOS vector is estimated from the position measurements in terms of pixel (as shown in Fig. 4.18) and is an input for a fuzzy controller. The controller outputs the rotation rate required to align the pointing vector along the LOS vector.

The main problem in the design of this fuzzy controller was a scaling factor choice. This was because the universe of discourse of the output strongly depended on the position of the object and its velocity in respect to the camera. If the object is very far away and has a typical approach velocity, its LOS vector is going to rotate slowly. If the object is close enough, its approach velocity may rotate the LOS vector with a high rate. In that case the object may go out of the FOV quickly.

To solve this problem, the decision was taken to evaluate the scaling factor depending on the distance. This problem was solved by the first part of the controller (Fig. 4.19) which outputs a fuzzy estimation of the scaling factor. This estimation is applied to the main part of the fuzzy controller.

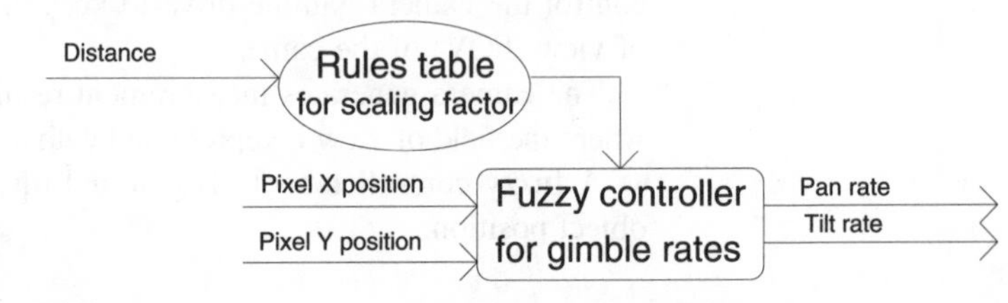

Fig. 4.19 *The structure of a fuzzy control system for a camera tracking*

One needs to remember that while changing the scaling factors, one actually modifies a linguistic sense of the inputs and outputs of the fuzzy controller. If the only purpose of this change is to improve a system performance without intending to maintain a linguistic essence of a fuzzy controller, then one may change the scaling factors. However, if the rules base has been formulated by the expert such an approach cannot be appropriate. Changing the scaling factors may result in losing the original linguistic sense of a rules base. The experts may not recognise their rules after tuning the scaling factors and will not be able to formulate new rules.

4.4 Membership function choice

4.4.1 Distributing membership functions on the universe of discourse

We concluded in the previous section that a fuzzy controller was similar to a linear PID controller. Is it right?

Yes it is. A fuzzy controller can be approximated as a PID-like controller very roughly. This approximation can be applied when one considers the problem of an initial scaling factors choice. In a case of a PID controller, a design problem includes a proper choice of the PID-controller coefficients. In a fuzzy controller design, one needs to choose many more parameters: membership functions, fuzzification and defuzzification procedures, etc. These extra parameters make a fuzzy controller more robust and much more difficult for analysis as well.

In the membership function choice, one has to solve a few problems: how to choose general parameters, such as the number of classes (membership functions) to describe all the values of the linguistic variable on the universe, the position of different membership functions on the universe of discourse, the width of the membership functions, and concrete parameters, such as the shape of a particular membership function.

We will try to provide general recommendations first of all. Let us start with the position of the classes on the universe. Suppose we have three classes: high, medium and low with the membership functions given in Fig. 4.20. What do you suggest here?

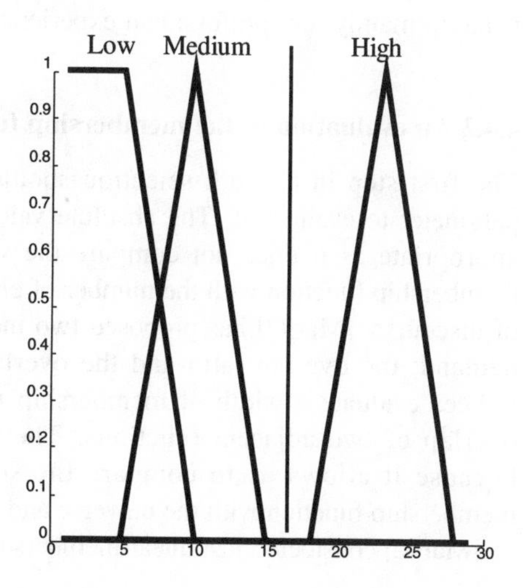

Fig. 4.20 *Membership functions (classes) distribution*

I guess it is not very good. If we take, for example, a real input signal of 16, it does not correspond to any class (the degree of matching any class is 0). It means that this real input does not fire any rule so the controller does not react on this input. If the membership functions distribution is supposed to describe output classes in some defuzzification methods, the area on which integration is performed will be discontinued.

I agree that the membership functions describing different classes should overlap. But how much? Which overlap is better, Fig. 4.21a or b? The choice of membership functions remains as the most undeveloped problem in fuzzy controller design.

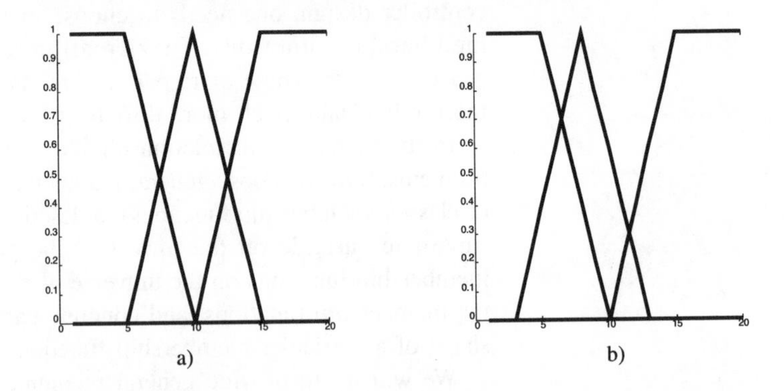

Fig. 4.21 *Membership functions distribution with different overlaps*

Software design packages propose different ways to define membership functions but give no, or in the best case very general, recommendations on how to select the membership function shape and width. Although some theoretical [Kre92] and practical [Fuz92] suggestions have been already made, this choice is based mainly on professional experience and common sense.

4.4.2 An evaluation of the membership function width

The first step in a width selection should be the choice of a parameter to evaluate it. The absolute value of the width is not appropriate as it does not compare the width of the separate membership function with the number of classes and the universe of discourse. [Mar92] has proposed two indices which meet this demand: the overlap ratio and the overlap robustness. These indices evaluate a width of membership functions through the overlap of two adjacent functions. The idea is very fruitful, because it allows us to compare the scope of the separate membership function with the universe and the number of classes.

[Mar92] considered just linear membership functions, and these indices worked very well. They can easily be calculated for all membership functions with a limited discourse area, however, this is not the case for functions with an infinite discourse area, for example, an exponential (Gaussian) membership function. Moreover, the first index does not depend on the membership function shape at all and the second one depends on the shape within the overlap area only.

For these reasons we would like to propose the index of the whole overlap:

$$WO = \frac{\int_x \min(\mu_1(x), \mu_2(x))}{\int_x \max(\mu_1(x), \mu_2(x))}$$

where $\mu_A(x)$ and $\mu_B(x)$ are two adjacent membership functions. The evaluation formulas for this index and some usual membership functions are given in Table 4.4.

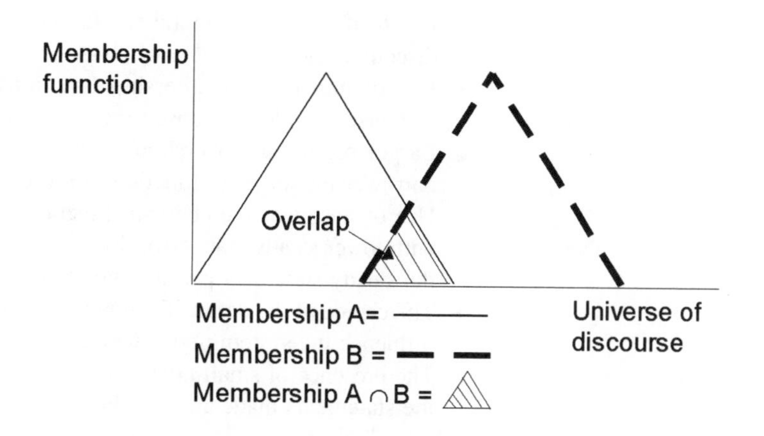

Fig. 4.22 *A whole overlap parameter evaluation*

Fine. But what do we need this parameter for?

We can see that the index of the whole overlap (WO) can easily be calculated in the cases of membership functions with the unlimited discourse area as well as in the cases of simple functions with a limited area. Also it gives us complete information about the width of a membership function, comparing it with a number of classes and an area of discourse. Judging all of these reasons, we can choose this parameter as a base to evaluate its influence and the influence of the membership functions position on the different performance indicators of a fuzzy controller.

Table 4.4 Whole overlap parameter for some membership functions

Membership function	Formula	WO ratio
Linear (triangular)	$\mu(x) = 1 - \text{abs}((x - m)/\sigma)$	$\dfrac{1}{\dfrac{2}{(1-\frac{m_2 - m_1}{2\sigma})^2} - 1}$
Exponential (Gaussian)	$\mu(x) = \exp[-((x - m)/\sigma)^2]$	$\dfrac{1}{F(\sqrt{2}\frac{m_2 - m_1}{2\sigma})} - 1$
Quadratic	$\mu(x) = \max(0, 1 - ((x - m)/\sigma)^2)$	$\dfrac{\frac{2}{3} - \frac{m_2 - m_1}{2\sigma} + \frac{1}{3}(\frac{m_2 - m_1}{2\sigma})^3}{\frac{2}{3} + \frac{m_2 - m_1}{2\sigma} - \frac{1}{3}(\frac{m_2 - m_1}{2\sigma})^3}$

The research results [Rez93] let us make the following assertions:

- The parameter of the whole overlap is appropriate for an evaluation of a membership function width as it compares the width of a separate membership function with the area of discourse and the number of classes.
- Use of narrower membership functions results in a faster response (smaller response time).
- Larger oscillation, overshoot and settling time appear when narrower membership functions are used.
- Use of narrower membership functions produces the system with lower steady-state error. But with a very narrow function, the steady state may possibly not be reached at all.
- The choice of the defuzzification method does not significantly influence the system performance characteristics.
- The presence of small noise and disturbances generally keeps the statements made above valid.

If we compare an action of a fuzzy controller with the action of a conventional PID controller, we can state that when we decrease a membership function width (decrease the WO ratio), we increase the differential part, and in the opposite case we emphasise an integration performance of the controller. So we can see some analogy between a membership functions choice and a PID controller coefficients choice similar to that we established for a scaling factor choice.

Has large noise presence been considered in this research?

Yes, it has. Generally speaking in the presence of large

disturbances it would be better to apply a fuzzy controller with a higher overlap.

4.4.3 Application example

Example 4.4 A temperature control servo system

A control servo system keeps a constant temperature for a conventional water heater tank. The water heater basically has five outputs: the temperature error (the difference between the desired temperature and the real one) and four other states related to the water level of the tank itself. Initially only the temperature was examined. If the temperature is too high, then the amount of necessary heat is less than the current amount. If it is too low, then more heat is necessary. The fuzzy controller with different membership functions has been used to control this plant. In Fig. 4.23 step responses of this control system with different overlaps are given. The Gaussian membership functions and means of maximum defuzzification methods were applied.

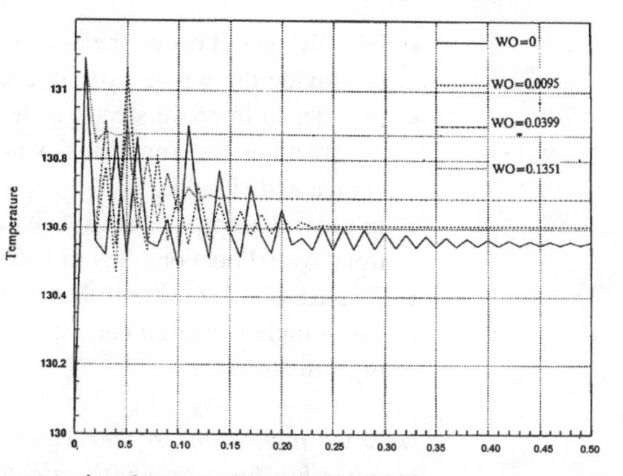

Fig. 4.23 *The step response for the tank temperature control system with various overlaps*

This was a simple example and even here one can see that such response parameters as the steady-state error, overshoot, settling time do depend on the overlap and consequently on the membership function choice. A more complicated one is our vessel control.

Example 4.5 A vessel movement control system

Figure 4.24 presents two different trajectories realised by similar fuzzy controllers. The only difference was in the overlap of the controller input and output membership functions. You see that the results obtained are quite different.

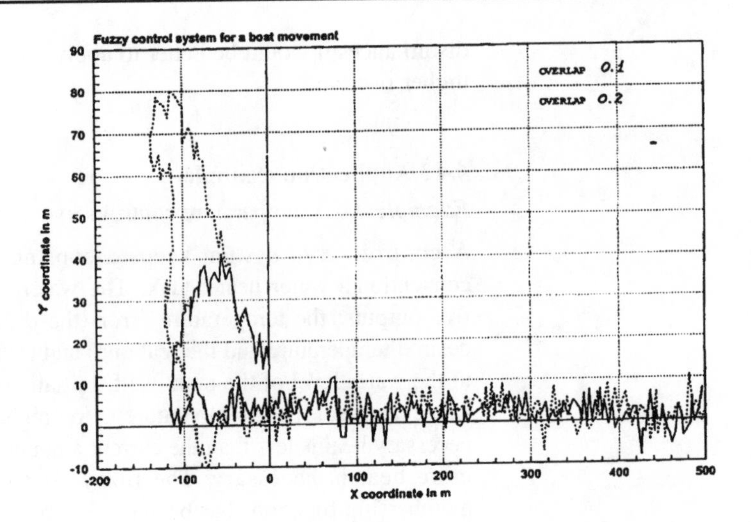

Fig. 4.24 *The various trajectories corresponding to different overlap*

And that is not all. On the base of this analysis a user can be advised as follows:

- Initially one chooses the width of the membership functions to provide the whole overlap about 12–14 per cent.
- In order to improve steady-state accuracy one has to decrease the membership functions' whole overlap after their initial choice and simulation.
- In order to improve stability characteristics (oscillation, settling time, overshoot) one has to increase the whole overlap.
- The use of a fuzzy controller with wider membership functions and a higher overlap can be recommended in the presence of high disturbances.

Now we have some rather clear recommendations about the membership function width selection and further modification. What about the choice of the shape?

Research results do not demonstrate a significant dependence of the performance indices on the shape of the membership functions generally. However, this dependence should be investigated in any particular case because some improvement can be achieved.

4.5 Fuzzy rule formulation

4.5.1 Where do rules come from?

It is very important to emphasise the role of the rules in a fuzzy controller. Some research has been conducted and some results have been achieved. If one tries to classify the approaches applied to the rules derivation [Lee90], the results are as Table 4.5 and Section 8.7.

Table 4.5 Main approaches and sources of fuzzy controller rules
Expert experience and knowledge.
Operator's control actions learning.
Fuzzy model of the process or object under control usage.
Learning technique application.

The first approach comes from an expert systems domain. The reason for that is pretty obvious: any fuzzy controller is actually an expert system applied to control problems. All theoretical and practical methods of knowledge acquisition developed in artificial intelligence and other sciences are to be applied here. It should be noted that by using linguistic variables, fuzzy rules provide a natural framework for human thinking and knowledge formulation. Many experts find that fuzzy control rules provide a convenient way to express their domain knowledge. So cooperation with the experts will be easier for a knowledge engineer.

How do we find the rules practically? I mean how should this process be organised?

Two methods can be proposed based on work with the documentation or the experts themselves. The first one is based on redeveloping manuals, operation instructions and any other documents available into the set of the rules. Another way includes an interrogation of experienced experts or operators using a carefully organised questionnaire. Of course, a good knowledge engineer always tries to apply the combination of these two approaches.

Another method is pretty similar to the first one. Here the rules are formulated by observing how a skilled operator controls the object or the process. In this case the operator should be an expert. As was pointed by Sugeno, in order to automate a control process, one can express the operator's control rules as fuzzy if–then rules employing linguistic variables. In practice, such rules can be deduced from observation of the human controller's actions.

Both of the approaches proposed seem rather subjective, depending on an expert opinion. Is there another method based on for example, some measured data?

Yes, there are two other methods. The first one is the generation of a set of fuzzy control rules based on the fuzzy model of the process or object under control. This approach is ideologically similar to a classical controller design technique in which we derive a mathematical model of the controller from the mathematical model of the plant. Here the mathematical plant model is fuzzy.

Another method is based on learning and experience gained. This method supposes employing adaptive and/or self-organising fuzzy controllers and is to be considered in Section 5 in greater detail.

All of these methods suppose a heuristic (at least initially) derivation of the fuzzy control rules. Obviously, there is a need in the following analysis and justification of the rules.

I understand that the rules are formulated on the base of some knowledge about the process, object or some data available. What can we do if almost nothing is available?

Although any special knowledge is very valuable in rules formulation, do not underestimate the 'classical' examples of the PID-like fuzzy controllers' rules. This can be a good starting point for many cases, especially with the analysis and adjustment to follow.

4.5.2 How do we get rules?

Example 4.6 Fuzzy controller for automated delivery of muscle relaxants [Mas94]

Anaesthetists are familiar with administering bolus injections of nondepolarising muscle relaxants at more or less regular intervals with minimal feedback as to the level of muscle relaxation. Computer controllers are needed nowadays to regulate the continuous infusion of muscle relaxant to match each individual patient's needs, avoiding overdose and maintaining stable relaxation. A fuzzy self-organising controller was proposed in [Mas94] for this problem. As there was no expertise available to build the rule base, because anaesthetists are not generally accustomed to controlling muscle relaxation by a continuous infusion, a general technique was applied. A PD-like fuzzy controller with two inputs was proposed: inputs were error and change of error, with the error being the deviation from the desired level of muscle relaxation. The output of the controller was the infusion rate. This controller has the following rule base (Table 4.6). One can see that a standard PD-like fuzzy controller was applied here because of the absence of special knowledge.

Table 4.6 Rule base for a fuzzy controller for automated delivery of muscle relaxants

e \ Δe	**NB**	**NM**	**NS**	**Z**	**PS**	**PM**	**PB**
NB	NB	NB	NB	NM	NM	NS	Z
NM	NB	NB	NM	NM	NS	Z	PS
NS	NB	NB	NS	NS	Z	PS	PM
NO	NB	NM	NS	Z	Z	PM	PB
PO	NB	NM	Z	Z	PS	PM	PB
PS	NM	NS	Z	PS	PS	PB	PB
PM	NS	Z	PS	PM	PM	PB	PB
PB	Z	PS	PM	PM	PB	PB	PB

4.5.3 How do we check if the rules are OK?

Usually the rules are formulated one by one. After this, one has to analyse the whole set of the rules. In this analysis, one needs to determine if the set of the rules is:

- complete;
- consistent;
- continuous.

How do we do that?

Let me start with the definitions. The rules set is complete if any combination of input values results in an appropriate output value. Any combination of inputs should fire at least one rule.

Example 4.7 The fuzzy controller has the following rules set (Table 4.7).

Table 4.7

e \ Δe	**NB**	**NS**	**Z**	**PS**	**PB**
NB	^{11}	^{12}	PB^{13}	PS^{14}	Z^{15}
NS	^{21}	PB^{22}	PS^{23}	PS^{24}	Z^{25}
Z	PS^{31}	PS^{32}	Z^{33}	Z^{34}	NS^{35}
PS	Z^{41}	Z^{42}	NS^{43}	NS^{44}	NB^{45}
PB	NS^{51}	NS^{52}	NB^{53}	NB^{54}	NB^{55}

Is this rules set complete?

I do not think it is, as there are some empty cells in the table. It means that some combinations of the inputs: (NB,NB), (NS,NB) and (NB,NS) will fire no rules. It will confuse the controller and produce no output in this case.

I agree that an incomplete rules set is not good. However, I reckon that empty cells do not necessarily mean that the set is incomplete.

But the definition says that any combination of the inputs must be possible.

Yes, but you forget that inputs of the fuzzy controller are crisp, and inputs of the rules table are fuzzy or fuzzified crisp. So we need to consider any combination of the crisp inputs and the membership functions for both inputs. For example, if we have the membership functions for the input classes like in Fig. 4.25, the rules set is complete.

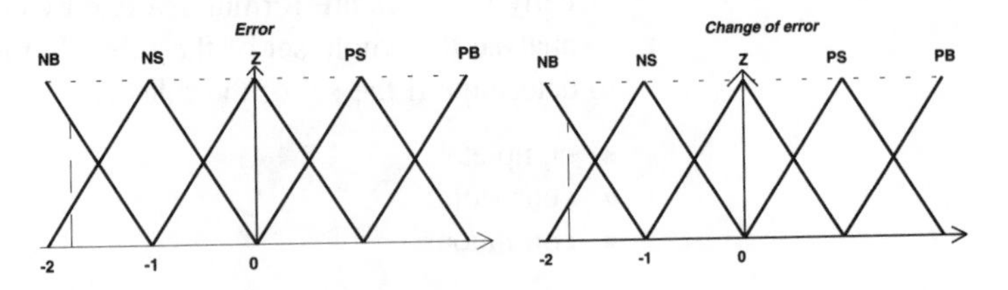

Fig. 4.25 *Membership functions for the inputs*

How come?

Let us consider two crisp inputs, both −1.8, which we suspect as having a very good chance of producing no outputs. After fuzzification, both of them will take a fuzzy value of (NB, 0.8) and (NS, 0.2). It means that to decide which rules are fired, we need to consider the combinations (NB,NB), (NB,NS), (NS,NB) and (NS,NS). The first three do not produce any output, but the fourth does. So this combination of the crisp inputs will fire one rule. Actually, in this controller, there is no combination of the *crisp* inputs which results in no outputs. Hence, this set is complete.

So the rules table may have empty cells.

Absolutely! And practically implemented fuzzy controllers usually have a few. It saves some memory and makes a controller operation faster.

A set of rules is consistent if it does not contain contradictions. By that I mean a situation when two similar antecedent parts produce different consequent parts, for example:

if error is Z **and** change of error is Z **then** output is Z

and:

if error is Z **and** change of error is Z **then** output is NB.

This is an obvious mistake and one of the rules must be excluded.

A set of rules is continuous if it does not have neighbouring rules with output fuzzy sets that have an empty intersection.

Hang on. Which rules are neighbouring? And what is this intersection?

We can define 'neighbouring rules' as follows: two rules are neighbours if their cells are neighbours. In the previous rules table, Rule 33 has neighbouring Rules 23, 32, 34, and 43. An intersection means that the rules do not allow a jump in output value for a small change in an input value. We have decided that a crossing point between two adjacent membership functions should not be zero. It means that this property requires a change to the adjacent class in any input to result in a change to an adjacent class in the output. If this small change in an input results in jump over a neighbouring class, the rules set is discontinuous.

What is wrong with this set?

In some situations, it could not provide smooth control. Sometimes it may significantly decrease the controller performance.

Rules can be formulated on a base of not just professional experience but a common knowledge and even common sense. An antecedent part of any rule may contain non-measured inputs which can be determined as fuzzy inputs initially, although this is not widespread in control applications. For example, in the problem of colouring a map [Ter94] the following rules are given:

if season is spring **then** colour is rather light green
if season is winter **then** colour is slightly grey brown.

4.5.4 Application examples

Example 4.8

Let us come back to our vessel control again and consider the rules sets in Tables 4.8 and 4.9.

Table 4.8

e \ Δe	**NB**	**NS**	**Z**	**PS**	**PB**
NB	PB	PB	PB	PS	Z
NS	PB	PB	PS	Z	NS
Z	PB	PS	Z	NS	NB
PS	PS	Z	NS	NB	NB
PB	Z	NS	NB	NB	NB

Table 4.9

e \ Δe	**NB**	**NS**	**Z**	**PS**	**PB**
NB	PB	PB	PB	PB	Z
NS	PB	PB	Z	Z	NB
Z	PB	PB	Z	Z	NB
PS	PB	Z	NB	NB	NB
PB	Z	NB	NB	NB	NB

Now try to compare these tables.

I see that the second table represents the discontinuous set of the rules.

And a result of this you may see in Fig. 4.26. The trajectory is quiet different from the previous one.

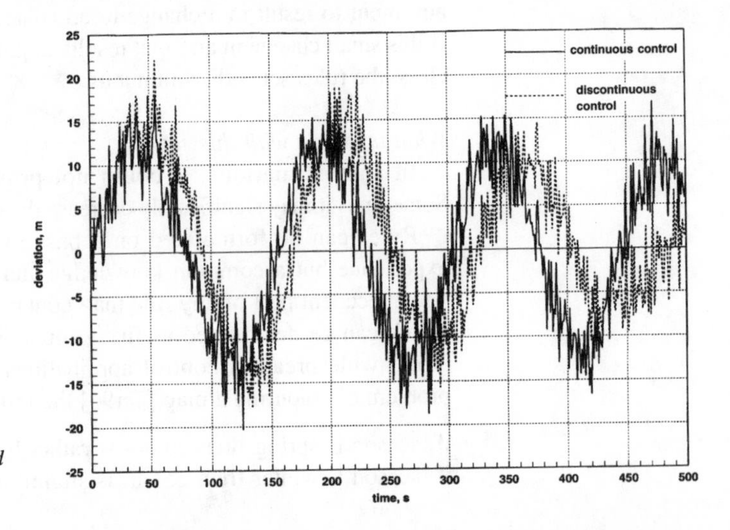

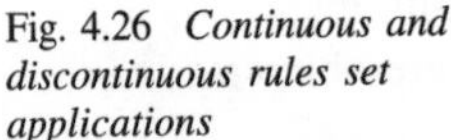

Fig. 4.26 *Continuous and discontinuous rules set applications*

Example 4.9 AC induction motor fuzzy controller [Clel92]

The most important challenge to reducing motor power consumption is to vary properly the shaft speed of motors which are designed as constant-speed machines. Electric motors use over 60 per cent of the electrical power generated in the United States[Bald87]. The efficiency of a constant-speed induction motor can drop drastically under reduced loads. To minimise power losses, it is necessary to control a motor speed and thereby adjust the motor speed to the load requirements. The most recent and successful approach is an adjustable speed drive (ASD). An embedded fuzzy controller can be added to conventional ASDs.

The fuzzy controller should solve the following problems:

- Assess a direction of change of an input power (P_{in}) to the motor and change the stator motor voltage V_s in a corresponding direction for reducing the input power.
- Sense when the input power was minimised to an extent that further variations in V_s produce negligible results.
- Control the step size for varying V_s so that convergence on the optimum operating point is accelerated.
- Limit perturbations to avoid an insufficient torque or excess speed (typical limits were ±5 per cent in respect of variations of the initially specified values).

Based on these goals, the rules table, describing how the stator voltage should be changed depending on how it and the input power were changed before, can be formulated as follows:

If ΔP_{in} is ***Neg*** **and** ΔV_{old} is ***Neg*** **then** ΔV_{new} is ***Neg***.
If ΔP_{in} is ***Neg*** **and** ΔV_{old} is ***Pos*** **then** ΔV_{new} is ***Pos***.
If ΔP_{in} is ***Pos*** **and** ΔV_{old} is ***Neg*** **then** ΔV_{new} is ***Pos***.
If ΔP_{in} is ***Pos*** **and** ΔV_{old} is ***Pos*** **then** ΔV_{new} is ***Neg***.
If ΔP_{in} is ***Zero*** **and** ΔV_{old} is ***Any*** **then** ΔV_{new} is ***Zero***.

The last rule is needed for the convergence on an optimal input power. However, this small set was found to be not sufficient to produce a good transient response. Additional linguistic variables (***Pos Medium*** and ***Neg Medium***) and rules were added to eliminate overshoot.

Example 4.10 Torque maximiser for a locomotive based on fuzzy rules [Hel94]

The actual driving force for a modern electrical locomotive is the friction force acting between a wheel and a rail. There should be a balance between the forces at the wheel and the motor shaft in which

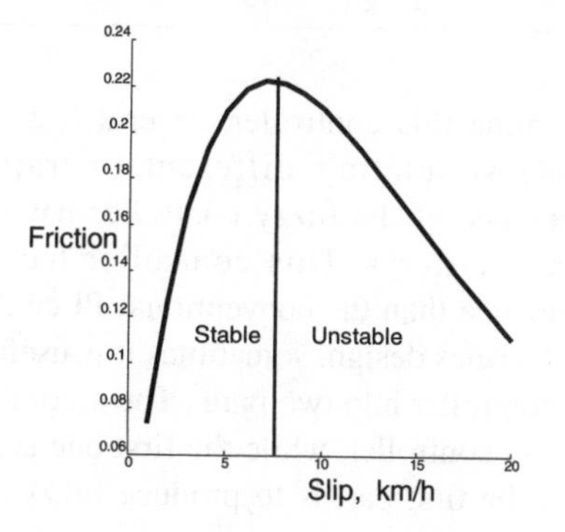

Fig. 4.27 *An electrical locomotive driving force as a function of a slip.*

a drive introduces a motor torque. The friction force depends on the slip and generally is a nonlinear function with parameters determined by the velocity of the locomotive and the quality of the wheel-rail conditions. Ideally the slip-friction curve should have a maximum or a plateau and can be divided into two regions: stable and unstable.

For the stable region (the rising branch of the curve) a non-controlled dynamic system vehicle-rail is always stable if the required motor torque in addition to the forces of inertia, damping forces and spring forces does not exceed the maximum of the friction force. However, if an increase of the motor torque disturbs the force balance, the system can be stabilised again. The aim of the torque maximiser is to keep the system characteristics close to the optimum point where the friction reaches the maximum. It is done by calculating the value of the slip correction (corr), which depends on the part of the curve the system is working at the moment. If the system reacts with a positive (or negative) change of the friction force Df, when a positive (or negative) slip correction is introduced, it means that the system is working on the stable branch of the curve. Otherwise, the system is in the unstable region. Because the closer the system is to the optimal point, the smoother the curve, a rough value for the correction can be determined on the base of the absolute values of the previous slip correction and force reaction. This description allows to formulate 25 rules in the table.

Table 4.10

Δs \ Δf	NB	NS	Z	PS	PB
NB	PB	PS	Z	NS	NB
NS	PS	PS	Z	NS	NS
Z	Z	Z	Z	Z	Z
PS	NS	NS	Z	PS	PS
PB	NB	NS	Z	PS	PB

To examine this controller, several tests have been performed [Hel94] simulating different operating conditions. The performance of the fuzzy controller has been evaluated against different criteria. This controller has demonstrated better performance than the conventional PI controller.

In the rules design, sometimes it is useful to subdivide a whole fuzzy controller into two parts. The second part can be considered as a main controller, while the first one is a preparation part. The goal of the first part is to produce fuzzy signals which are used in the rules table of the second part. In this way, the preparation part can be applied to solve classification problems.

Example 4.11 Automotive automatic transmission fuzzy controller [Hir93]

This fuzzy controller is used for a shift scheduling of automatic transmission vehicles. Usually the vehicle velocity and the throttle angle are the only two factors that are applied for determination of the shift scheduling in an automatic transmission. The influence of these factors can be expressed with the following rules:

If speed is ***low*** **and** throttle is ***low*** **then** shift **+1**.
If speed is ***low*** **and** throttle is ***high*** **then** shift **–3**.
If speed is ***high*** **and** throttle is ***low*** **then** shift **+3**.
If speed is ***high*** **and** throttle is ***high*** **then** shift **–1**.

By contrast, a driver using a manual transmission makes the gear shift decision based on a consideration of some extra factors, such as the road resistance, the current shift, the brake time, the vehicle acceleration. To improve a shift scheduling, one can introduce some extra rules describing the influence of these extra factors:

If resistance is ***NegBig*** **and** throttle is ***Close*** **and** shift is ***NotLow*** **then** shift **–2**.
If speed is ***Low*** **and** shift is ***High*** **then** shift is **–3**.
If brake time is ***Long*** **and** shift is ***NotLow*** **then** shift is **–2**.

An introduction of the extra rules makes the controller operation similar to a human driver. Considering these rules in combination with the previous set, one can significantly improve a performance of the controller, especially in uphill driving.

Example 4.12 Siemens automatic transmission control [Hel94]

Let us consider how this rules table can be generated in greater detail. Taking into account inputs from the accelerator and brake pedals, gearbox and other areas, and processing them with the help of two dozen rules, a fuzzy controller can deduce whether a vehicle is climbing or descending, and thus optimise vital motor functions to meet not only the driver's habits and style, but to enhance safety and reduce fuel consumption. The fuzzy control system can consist of two parts: the preparation part and the control. The first part receives seven input signals (e.g. throttle opening, *dki*, change in *dki*, etc.) and produces three output signals (Fig. 4.28): one is used for the load detection, one for the driver classification and one is a shift signal indicating whether this shift is allowed or not. Both the load and driver signals are used to choose a shift pattern.

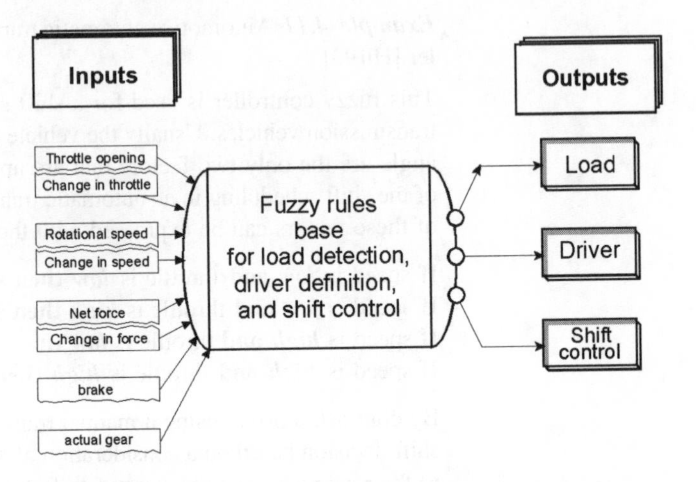

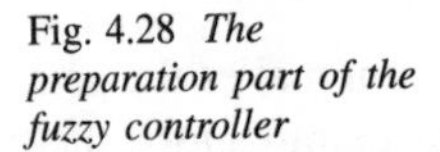
Fig. 4.28 *The preparation part of the fuzzy controller*

The importance of a shift pattern adaptation to the driver and load could be illustrated with the following example. In the case of an increasing load or a sporty driver, shift patterns with flat curve characteristics are selected. This leads to a delay in shifting up or even to shifting down. Therefore, the engine will run at a higher speed. The third signal (indicating which shift is allowed) limits possibilities, depending on the engine construction. For example, if there are four gear positions available, it is not possible to shift up when driving in the fourth gear. Shifting down must be forbidden when driving too fast for a lower gear position.

This part allows us to prepare conditions for a proper gear choice. The system includes some rules describing the behaviour of a normal driver, for example:

If *dki* is ***Very Big*** **and** n_{ab} is ***Not Very Big*** **and** Dn_{ab} is ***Not Positive*** **then** load is ***Mountain.***

Another set of rules allows us to take into account the typical behaviour of a driver. The driver classification depends on the change of throttle opening and the changes of it, for example:

If Δdki is ***High*** **and** Δn_{ab} is ***High*** **then** the driver is ***sporty***.

The shift signal has four discrete states: ***Shift Forbidden, Shift Allowed, Shift Down Forbidden, Shift Up Forbidden***. The rules including this input in their antecedent parts provide a proper technological system behaviour and a system security.

4.6 Choice of the defuzzification procedure

Here we give a very brief description of the most widely used defuzzification procedures and compare them. Let me remind you that the defuzzification goal in Mamdani-type fuzzy controllers is to produce a crisp output taking the fuzzy output obtained after rules processing, clipping or scaling (see Fig. 3.7). No defuzzification is necessary in a Sugeno-type fuzzy controller. Because the input for the defuzzification procedure is the clipped (or scaled) membership function, most of the methods propose the procedure of calculating the integrated evaluation of this input.

We will consider the following methods:

- Centre-of-area/gravity defuzzification.
- Centre-of-largest-area defuzzification.
- Mean of maxima.
- First-of-maxima defuzzification.
- Middle-of-maxima defuzzification.
- Height defuzzification.

Why are there so many methods?

I suppose the main reason is that none of them has proved its advantage over others. Nowadays the choice of the defuzzification procedure is based mainly on personal preference. It is even hard to introduce these methods, as different authors sometimes name the same procedure differently.

The first two methods tend to produce an integral output considering all the elements of the resulting fuzzy set with the corresponding weights. Other methods take into account just the elements corresponding to the maximum points of the resulting membership functions.

4.6.1 Centre-of-area/gravity

The centre-of-area method (in the literature also referred to as the centre-of-gravity method) is the most well-known defuzzification method. In the discrete case (when we have some values of the output signal with the corresponding membership degrees $\{u_1/\mu(u_1), ..., u_k/\mu(u_k)\}$) this results in

$$u^* = \frac{\sum_{i=1}^{k} u_i * \mu(u_i)}{\sum_{i=1}^{k} \mu(u_i)}$$

In the continuous case we obtain

$$u^* = \frac{\int_U u * \mu(u)\, du}{\int_U \mu(u)\, du}$$

where ∫ is the classical integral and μ (u) is a combined membership function.

This method determines the centre of the area below the combined membership function (Fig. 4.29).

This method sometimes includes a threshold or an index element T. Its application eliminates in the evaluation of u* those elements u_i where the membership degree is less than T. This threshold is selected at a value that reflects disturbances in both input data and properties of the object under control. In this case, the result is the centre of the area between the threshold and the combined membership function.

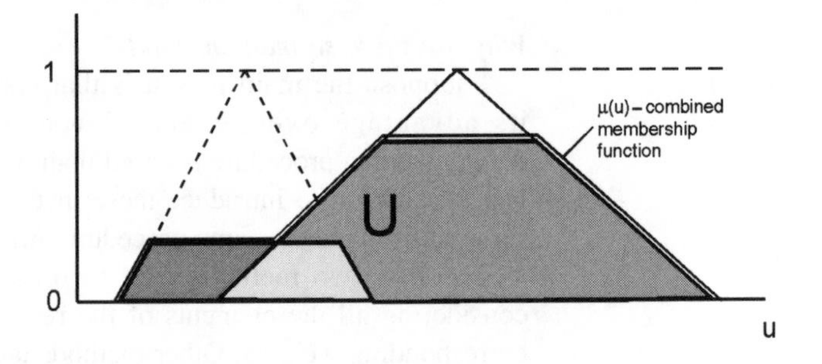

Fig. 4.29 *Centre-of-area defuzzification method*

It can be seen that this defuzzification method takes into account the area of U as a whole. Thus, if the areas of two clipped fuzzy sets are overlapped, then the overlapping area is not paid any particular attention. Because of integration, this operation is computationally rather complex and therefore results in quite slow inference cycles.

4.6.2 Centre-of-largest-area

The centre-of-largest-area is used in the case when U is non-convex, i.e. it consists of at least two convex fuzzy subsets which are not overlapped. Then the method determines the convex fuzzy subset with the largest area and defines the crisp output value u* to be the centre-of-area of this particular fuzzy subset (Fig. 4.30). It is difficult to represent this defuzzification method formally, because it involves first finding the convex fuzzy subsets, then computing their areas, etc.

In this method the defuzzification result is biased towards a side of one membership function. It does not take into account other parts of the clipped (scaled) membership function. Why not take an integral characteristic as produced by the centre-of-area method?

You are right, one can consider the centre-of-area estimation. However, in that case the defuzzification result (the crisp output) may be equal to the element with a very small or even zero degree. That estimation is not plausible as we will discuss a little bit later. To satisfy the criterion of plausibility, this method has been proposed.

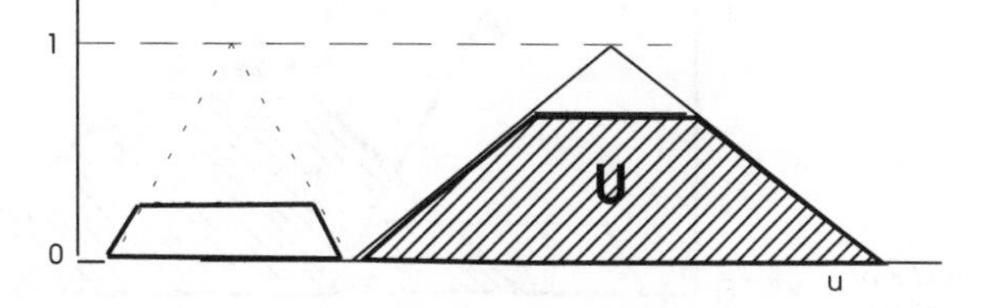

Fig. 4.30 *Centre-of-largest area defuzzification procedure*

4.6.3 First-of-maxima/last-of-maxima

First-of-maxima uses U and takes the smallest value of the domain U with the maximal membership degree in U. This is realised formally in three steps:

Let $hgt(U) = \sup_{u \in U} \mu_U(u)$ be the highest membership degree of U, and let

$\{u \in U \mid \mu_U(u) = hgt(U)\}$ be the set of domain elements with degree of membership equal to hgt(U).

Then u* is given by $u^* = \inf_{u \in U} \{u \in U | \mu_U(u) = hgt(u)\}$

The alternative version of this method is called last-of-maxima and is given as

$$u^* = \sup_{u \in U} \{u \in U | \mu_U(u) = hgt(u)\}$$

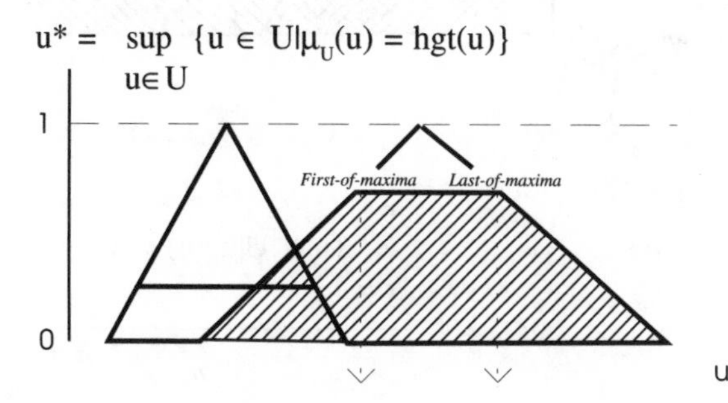

Fig. 4.31 *First-of-maxima/last-of-maxima defuzzification procedure*

4.6.4 Middle-of-maxima

Middle-of-maxima is very similar to first-of-maxima or last-of-maxima. Instead of determining u* to be the first or last from all values where U has the maximal membership degree, this method takes the average of these two values (see Fig. 4.32). Formally it can be written as

$$u^* = \frac{\inf\limits_{u \in U} \{u \in U | \mu_U(u) = hgt(u)\} + \sup\limits_{u \in U} \{u \in U | \mu_U(u) = hgt(u)\}}{2}$$

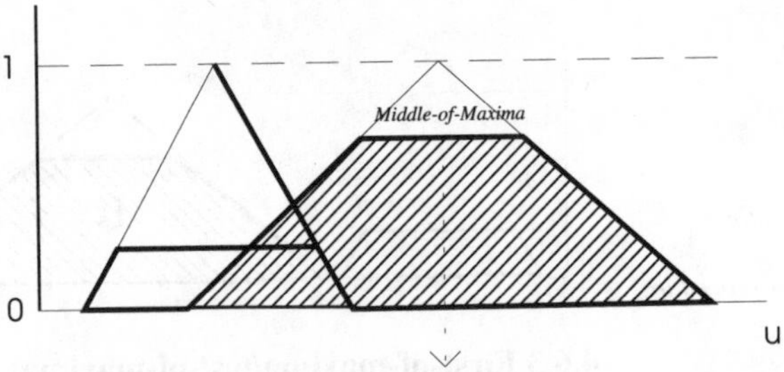

Fig. 4.32 *Middle-of-maxima defuzzification procedure*

4.6.5 Mean-of-maxima

This method does not differentiate between the elements of the combined membership function, instead it considers all of them and takes the average. If the membership function has *n* maximal points then the output will be:

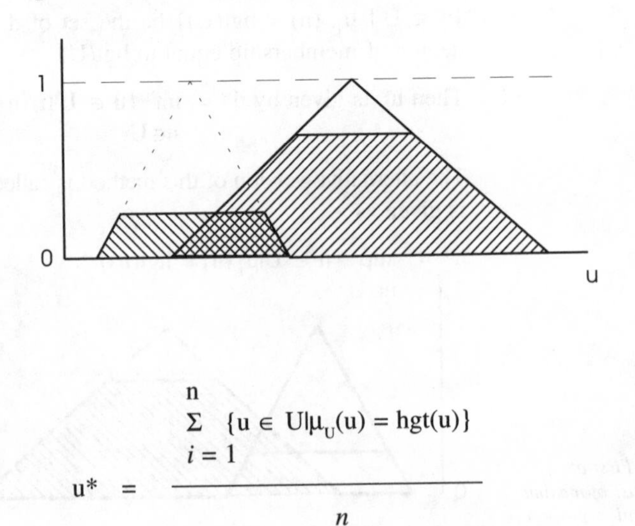

Fig. 4.33 *Mean-of-maxima defuzzification procedure*

$$u^* = \frac{\sum\limits_{i=1}^{n} \{u \in U | \mu_U(u) = hgt(u)\}}{n}$$

4.6.6 Height

Height defuzzification is a method which, instead of using U, applies the individual clipped or scaled control outputs. This method is very similar to the previous one. It takes the peak value of each clipped membership function CLU and builds the weighted (with respect to the height fk of CLU) sum of these peak values. Thus, neither the support or shape of CLU play a role in the computation of u*.

4.6.7 Compare different defuzzification procedures

I understand there are a lot of different procedures. Could you recommend any criteria for choosing a particular one?

I cannot recommend one criteria, but you may apply a few. Some criteria for a defuzzification method choice could be:

- continuity;
- disambiguity;
- plausibility;
- computational complexity.

Continuity again?

That's right. Now it is in respect to a defuzzification procedure. It means that a small change in an input of a fuzzy controller should not result in a large change in the output

Disambiguity means that a defuzzification method should work in any situation. This is not always the case. For example, the centre-of-largest-area method cannot make a choice when we have two equal areas.

Plausibility means that a procedure produces a defuzzified output that lies approximately in the middle of the support of the resulting membership function and has a high degree of membership. The centre-of-area method, for example, does not satisfy these properties: although the centre-of-area output lies in the middle of the support set, its membership degree may be one of the lowest possible.

The computational complexity criterion is particularly important in practical applications of fuzzy controllers. The methods dealing with the maximal points are fast methods, whereas the centre-of-area method is slower. The computational complexity sometimes depends on the shape of the output membership functions and whether max–min composition-based inference or scaled inference is chosen. The middle-of-maxima method, for example, is faster with a scaled inference. There is also an issue of representation of the fuzzy sets. In the case of the centre-of-area defuzzification, a tabular representation of the fuzzy sets in combination with clipped fuzzy sets (max–min composition) makes the defuzzification procedure very slow.

Let us summarise the features of all the methods in Table 4.11 [Drian93] where:

- CoA is centre-of-area;
- MoM is middle-of-maxima;
- FoM is first-of-maxima;
- HM is height and mean-of-maxima method;
- CLA is centre-of-largest-area,
- No* is only in the case of scaled inference.

Table 4.11

	CoA	MoM	FoM	HM	CLA
Continuity	Yes	No	No	Yes	No
Disambiguity	Yes	Yes	Yes	Yes	No
Plausibility	Yes	No*	No	Yes	Yes
Computational complexity	Bad	Good	Good	Good	Bad

You can use this table to make your choice depending on your preference.

5 FUZZY CONTROLLER PARAMETER ADJUSTMENT

5.1 Self-organising, adaptive and learning fuzzy controllers: the main principles and methods

5.1.1 What do we need adjustments for?

What do we need this second stage for? When we design a conventional controller, we do not adjust it. We just derive a mathematical model of the controller and then implement it.

Yes, you are right. In many conventional cases, we do not tune the parameters of a conventional controller. In those cases we know the exact model of the plant or object under control and, based on it, we are able to develop a mathematical model for the controller. The necessity for applying further tuning arises in situations where a controller must operate under conditions of uncertainty, and when the available *a priori* information is so limited that it is impossible or impractical to design in advance a controller that has fixed properties and also performs sufficiently well [Tsyp73]. You know that operational uncertainty is quite typical in a fuzzy controller. On the other hand, if conventional design methods suppose using a plant model, known *a priori* (before the design starts), in fuzzy control one usually has very limited, if any, knowledge about the plant. Even if the mathematical model is available, it usually includes some elements (nonlinearity, uncertainty) which make application of classical methods impossible. Moreover, a fuzzy approach assumes a design carried out under conditions of uncertainty. In this context, tuning can be viewed as a means for solving those problems that lack sufficient *a priori* information to allow a complete and fixed control system design to be achieved in advance.

But can we design a good fuzzy controller at the first stage?

Of course you can. In some cases (usually with simple problems), the controller's design solves the problems, demonstrates a satisfactory performance, and does not need any further reconsideration. The goal of the second stage is to enable a wider class of problems to be solved by reducing the prior uncertainty to the point where satisfactory solutions can be obtained on-line.

***A priori* design information available**

GOOD	➡	Fixed control design (1st stage only)
POOR	➡	Adjustment and tuning necessary (1st and 2nd stages)

Fig. 5.1 *Is the second stage necessary?*

I guess another good reason for tuning is changes in the plant and the environment in time. When the operational conditions are modified, a designer has to adjust the controller.

Right.

5.1.2 Self-organising fuzzy controllers

How do I tune the parameters of a fuzzy controller? I mean, how do I decide which parameters to change and by how much?

These criteria should be based on some evaluation of what is good and what is bad. Usually one or a few of the parameters characterising the controller performance are applied as a criterion. If the performance indicator is improved as a result of tuning, this adjustment should be accepted.

So the controller must not control but evaluate itself, right?

Not exactly. The controller has to control and evaluate the control performance.

But the controller structure does not suppose such evaluation, does it?

It means that the controller structure of Fig. 3.10 must be replaced with another one, see Fig. 5.2. We have added the metalevel of the controller which evaluates a performance and changes parameters. This structure represents a so-called self-organising fuzzy controller. The first controller of this type was proposed by Mamdani [Mam79].

5.1.3 Performance/robustness problem and solutions

The main goal of this self-organisation feature is to make the controller robust as much as possible to any changes. A trade-off

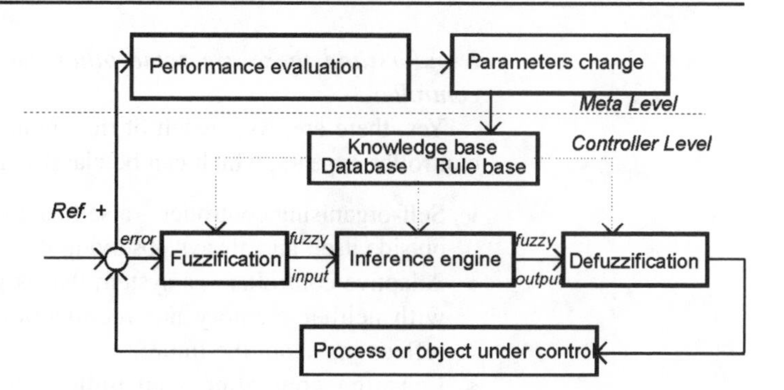

Fig. 5.2 *The self-organising fuzzy logic controller*

exists between performance and robustness, since robust control designs are usually achieved at the expense of the resulting closed-loop system performance (relative to a control design based on a perfect model). Advanced robust control system design methods have been developed to minimise this inherent performance/robustness trade-off.

Although robust design methods are currently limited to linear problems, nonlinear problems with model uncertainty can sometimes be approached by gain scheduling, a representative set of robust point designs over the full operating envelope of the plant, thus decreasing the amount of model uncertainty that each linear point design must accommodate. Nevertheless, the performance resulting from any fixed control design is always limited by the availability and accuracy of *a priori* design information.

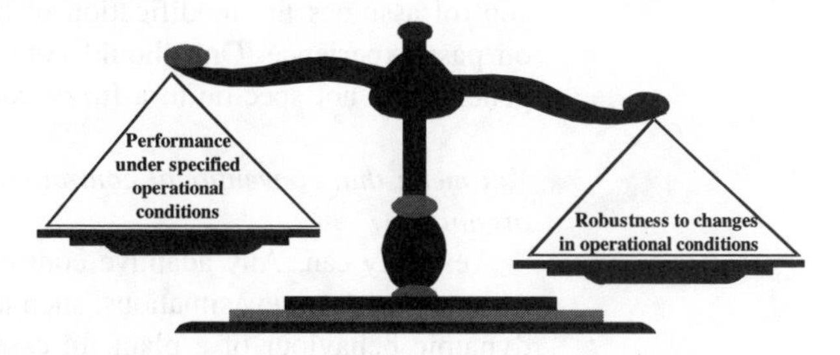

Fig. 5.3 *The trade-off between performance and robustness is a typical design problem*

If there is sufficient complexity or uncertainty so that a fixed control design will not suffice, then a satisfactory closed-loop system performance can only be obtained in one of three ways [Far92]:

- improved modelling to reduce the uncertainty;
- via an automatic on-line adjustment technique;
- manual tuning of the nominal control law design.

I understand there are some other ways for tuning a fuzzy controller.

Yes, there are. As a result of this tuning process, a new fuzzy controller appears, which can be classified as:

- Self-organising controller – a selfish creation noticing nothing outside itself and always observing itself only and nothing else.
- Adaptive controller – a system that is just a current situation with neither memory nor recollections about the past and reflections about the future.
- Learning controller – an industrious student constantly developing and expanding his or her knowledge and experience.

Although this classification is not contentious, the definitions are not easy to formulate. There are some different opinions because of changing understanding; however, here we try to present the most common point of view.

5.1.4 Adaptive fuzzy controllers

What is the difference between a self-organising and an adaptive controller?

A self-organising controller can be determined as a controller, the parameters of which are changed without any other changes in the plant model and other models used. An adaptive control usually supposes modifications of the plant model, and a learning control assumes the modification of the control strategy based on past experience. One should note that these definitions are general and not specific to a fuzzy controller.

You mean that conventional controllers can be adaptive or self-organising?

Yes, they can. Any adaptive control system can adjust itself to accommodate new situations, such as changes in the observed dynamic behaviour of a plant. In essence, adaptive techniques monitor the input–output behaviour of the plant to identify, either explicitly or implicitly, the parameters of an assumed dynamic model. The control system parameters are then adjusted to achieve some desired performance objectives. Thus, adaptive techniques seek to achieve an increased performance by updating or refining some representation, which is determined (in whole or in part) by a model of the plant structure, based on an on-line measurement information. An adaptive control system will attempt to adapt, if the behaviour of the plant changes by a significant degree. The structure for an adaptive fuzzy controller is given in Fig. 5.4.

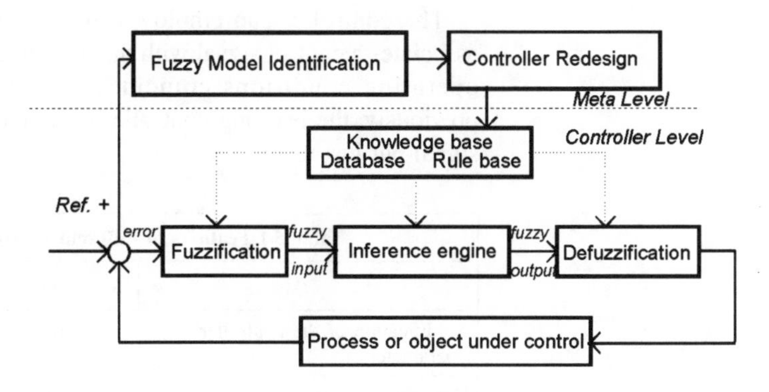

Fig. 5.4 *Adaptive fuzzy logic controller*

What do we need any other type of tuning for?

The problem is time for adaptation. If the dynamic characteristics of the plant vary considerably over its operation (e.g., due to nonlinear dynamics), then the control system may be required to adapt continually. During these adaptation periods, a high closed-loop performance cannot be guaranteed. Note that this adaptation can occur even in the absence of time-varying dynamics and disturbances, since the control system must readapt *every* time a different dynamic regime is encountered (i.e., one for which the current control law is inadequate), even if it is returning to an operating condition it has encountered and successfully adapted to before. So the adaptation must take place in every case of changing. This inefficiency results in degraded performance, since the transient behaviour due to the parameter adjustment will occur every time the recently observed dynamic behaviour of the plant changes by a sufficient degree.

In general, adaptive controllers operate by optimising a small set of adjustable parameters to account for a plant behaviour that is local in both state-space and time. To be effective, adaptive controllers must have relatively fast dynamics so that they can react quickly to any change of the plant behaviour. In some instances, the plant parameters may vary so fast (perhaps due to nonlinearity), that the adaptive system cannot maintain desired performance through an adaptive action alone. As argued by [Fu64], it is in this type of situation where a learning system is preferable. Because the learning system retains some information, or we may say, some knowledge, it can in principle react more rapidly to spatial variations once it has learned.

5.1.5 Features of different controller types

How can the reaction time be shortened?

The controller can employ a mechanism of learning which associates a control signal with the operating conditions. So if the operating conditions coincide with the ones experienced previously, the learning controller does not calculate the control signal but just restores it.

Table 5.1 Features of different controller types

	Self-organising	Adaptive	Learning
Modification of the controller parameters	Yes	Yes	Yes
Controller redesign	No	Yes	Maybe
Limitations in redesign and updating	Very strong	Strong	No
Identification and modification of the plant model	No	Yes	Maybe
Ability to accommodate novel operating environment	Poor	Good	Poor
Time for adaptation	Medium	Long	Short
Ability to control slowly changing with time processes	Poor	Good	Poor
Ability to control the processes and objects with unknown models	Very good	Poor	Good
Memory requirement for implementation	No	No	Large
Accuracy of the control law derived	Poor	Good	Poor
Ability to preserve high performance during functioning	Poor	Medium	Good

This controller will need some memory to store these conditions–control associations. So do you pay for speed with memory?

Yes, you are absolutely right. This is another example of the usual speed–memory trade-off. However, due to this, the designer can find the best possible way to solve any particular problem. The processes of adaptation and learning are complementary: each has unique desirable characteristics. For example, adaptive capabilities can accommodate dynamics that vary slowly with time and novel situations (e.g., those which have never been experienced before), but are often inefficient for problems involving significant nonlinear dynamics. Learning approaches, in contrast, have the opposite characteristics: they are well equipped to accommodate poorly modelled nonlinear dynamic behaviour, but are not well suited to applications involving time varying dynamics.

5.1.6 Learning fuzzy controllers

You have not provided the structure of a learning fuzzy controller.

No, I have not. It is not easy to draw the structure of a learning fuzzy controller, but it can be classified into three groups [Lin95]:

- supervised learning;
- reinforcement learning;
- unsupervised learning.

In the first two regimes, the learning process is based on the feedback signal, indicating the performance of the whole system under current conditions. In supervised learning the desired objective is provided at each time step. The performance indicator can be and is applied for an evaluation of the degree in which the current set of parameters satisfies the formulated objective. This learning regime can be considered as similar to a self-organising controller. In reinforcement learning, the teacher's response is not immediate and direct and serves more to evaluate a state of the system. In this aspect this regime looks like an adaptive controller. In early learning control systems, the output of the objective function was mapped to a binary reinforcement signal. In such situations the reinforcement signals could be thought of as being 'positive' or 'negative'. Positive reinforcement was designed to encourage the same behaviour in future, when the same control situation appears, while negative reinforcement was designed to discourage that behaviour. Unsupervised learning does not require any feedback, because this can be inefficient in control applications, and is not applied often. There are different ways of how the tuning process of the fuzzy controllers can be realised. Conventional methods, e.g., gradient analysis can be applied for these purposes. In this way the performance criteria are analysed as functions of the parameters to be adjusted. The directions of change are determined on the base of this analysis.

We will consider three methods originating from intelligent technologies:

- tuning of the fuzzy controller parameters with a fuzzy decision-maker (controller);
- neural network tuning;
- genetic/evolutionary algorithms tuning

5.2 Tuning of the fuzzy controller scaling factors

5.2.1 On-line and off-line tuning

Scaling factors are very important parameters of a fuzzy controller. Their role and some general recommendations about their choice were considered in Section 4.3. Here we will try to give more detail and consider practical examples. Generally speaking when we change values of scaling factors between the runs (when the controller does not work or is not tested), that is called off-line tuning, and during the run, that is called on-line tuning. We will consider both cases. Both input and output scaling factors can be tuned. In the first example, just the output scaling factors are adjusted. In the second one, the input and output scaling factors are both adjusted.

5.2.2 Off-line tuning of the output scaling factors

Example 5.1 *Self-organising fuzzy controller for an electric motor.*

Let us consider the electric motor given by its transfer function

$$G(s) = \frac{8}{(s+1)(s+1.6)}$$

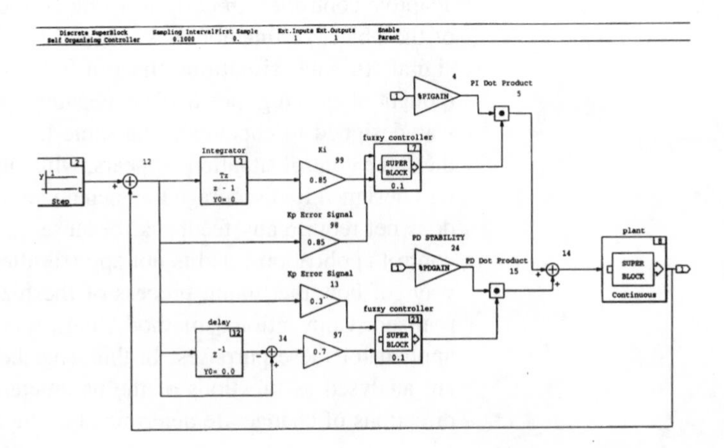

Fig. 5.5 *The MATRIXx™ model of the control system with a PID-like fuzzy compensator*

This plant was chosen as an object to control. The uncompensated closed loop system has characteristics as following:

- maximum overshoot is 22.08 per cent;
- ±5% settling time is 2.3077 s;
- unit step steady-state error is 0.888.

Obviously it badly needs a compensator. The fuzzy PID controller incorporating two parallel fuzzy controllers with the structure as in Fig. 3.15 is applied. The structure of the control system including the fuzzy PID (MATRIXx™ model) is given in Fig. 5.5. The output scaling factors of both fuzzy controllers should be adjusted.

Fine. How do we adjust these factors?

Some advice was given in section 4.3. We can try to formulate some rules like:

If settling time is large **and** overshoot is negative **then** increase K_p and K_I coefficients.
If settling time is large **and** overshoot is positive **then** decrease K_p and K_I coefficients.
If settling time is medium **then** increase K_p and K_{II} coefficients.
If overshoot is large **then** decrease K_p and K_I coefficients.

Implementing even such a simple set can improve your response significantly. Figure 5.6 demonstrates how the parameters of the response are changed when the output scaling factors are tuned.

How long does it last? I mean how many times should one run or simulate the controller to get an acceptable response and the corresponding values of the parameters?

The answer depends on the initial choice but not significantly. Figures 5.6–5.8 demonstrate how the response is changed when an initial PI gain is too small (Fig. 5.6) or large (Fig. 5.7) as well as when the initial choice is not bad (Fig. 5.8).

The rules given above do not refer to the integral error.

That is because such rules are not so easy to formulate. The more complicated structure is necessary in this case.

5.2.3 On-line tuning of the input and output scaling factors

On-line tuning assumes constant changing of the scaling factors. In many cases, it allows good results to be achieved much faster. The following example describes a system of simultaneous on-line tuning of both input and output scaling factors. It may look a bit complicated, but then the problem is not that simple.

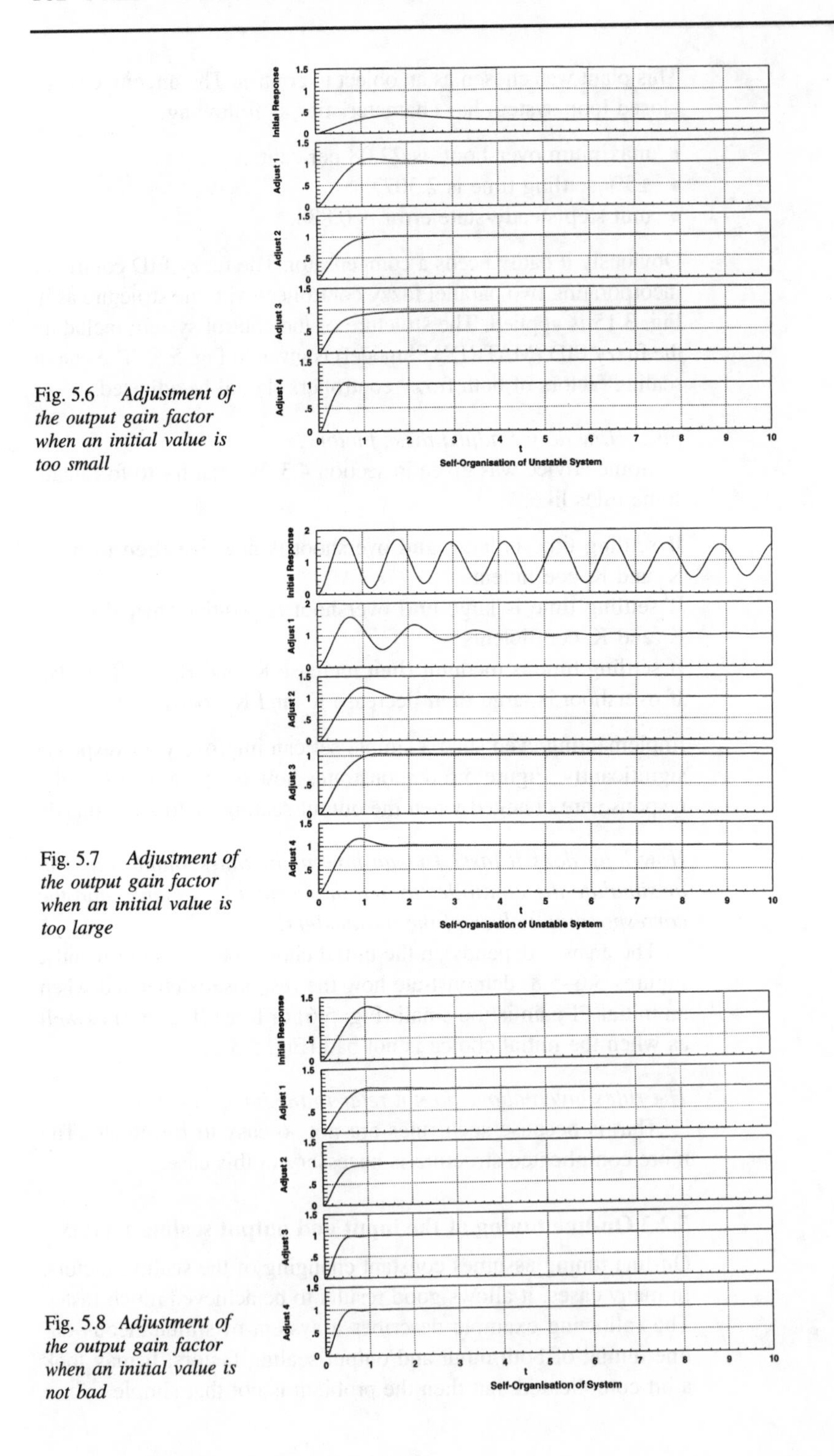

Fig. 5.6 *Adjustment of the output gain factor when an initial value is too small*

Fig. 5.7 *Adjustment of the output gain factor when an initial value is too large*

Fig. 5.8 *Adjustment of the output gain factor when an initial value is not bad*

5.2.4 Application example

***Example** 5.2* Self-organising fuzzy controller for a power system stabiliser (PSS)

The wide area of fuzzy controller applications in power system generation control involves an excitation system and a PSS design, and many successful applications have been reported [Hsu93]. A brief description of the problem and the simplest fuzzy controller applied was presented in Example 3.5. The most popular fuzzy controller structure and algorithm used in PSS design is one proposed by [Hiy89]. Here this method is advanced to achieve a multisectoral plane.

Let us try to find a solution as universal as possible. We want to design a universal tuner that can be used to adjust the set of the input and output scaling factors. This proposed calculator will operate together with a main fuzzy controller, i.e., the controller applied to manage the plant and designed in Section 3.6 and will change the scaling factors of the fuzzy controller. Any modification will be performed on-line. The purpose of the tuner is an on-line change of the fuzzy controller scaling factors in order to provide an acceptable system response. A general layout of the proposed system is shown in Figure 5.9. The proposed scheme requires some performance monitors calculated on the base of the current values of the main fuzzy controller signals. The error signal can be employed as such a monitor.

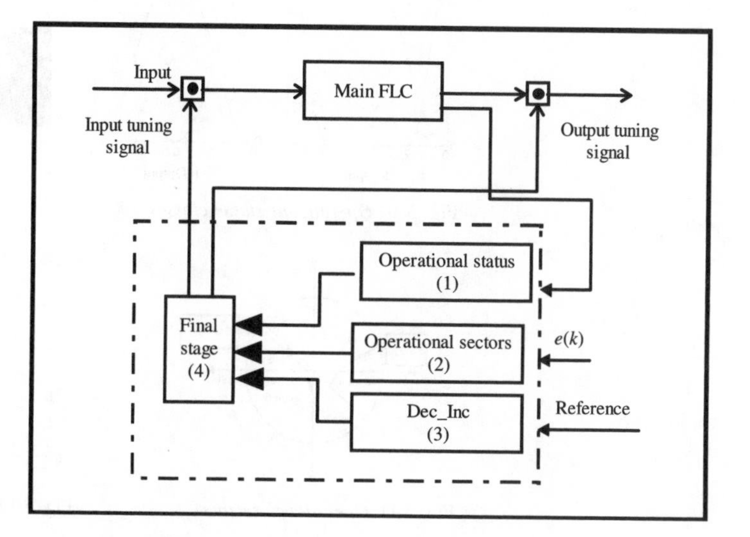

Fig. 5.9 *The tuner structure*

The four blocks labelled 1–4 in Fig. 5.9 are the main parts of the tuner. Blocks 2 and 3 employ the same principles as used in the Shay structure described in [Ghan95-2].

How does it work?

Block 1 considers the current state of the main fuzzy controller operational status. The upper and lower limits of the main controller response are of great importance in the operation of this block. This block applies fuzzy values of the output of the main fuzzy controller at two successive samples as two inputs ($fz(t)$ and $fz(t-1)$) and produces a fuzzy value representing the operational status. This fuzzy output may have three values:

- R_1 – under operation, the response currently is changing too slowly;
- R_2 – normal operation, the speed of change is acceptable;
- R_3 – over operation, the response is changing too fast.

Input classes of this block are shown in Fig. 5.10a, where classes R_{11} and R_{21} (R_1 of Input 1 and R_1 of Input 2) are limited by the lower class in the main fuzzy controller (magnitude). Classes R_{13} and R_{23} are limited by the upper limits of the main fuzzy controller. The output classes are shown in Fig. 5.10b and the fuzzy value of the operational status is calculated by the rule base in Table 5.2. Determining an operational status gives a sufficient indication of required tuning for either the input or the output ranges, or both.

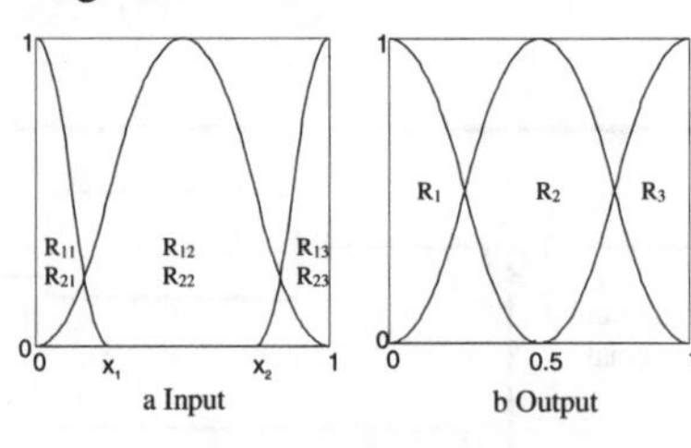

Fig. 5.10 *Operational status classes*

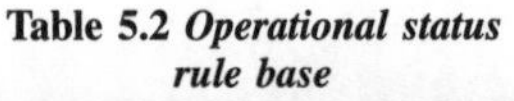

Table 5.2 ***Operational status rule base***

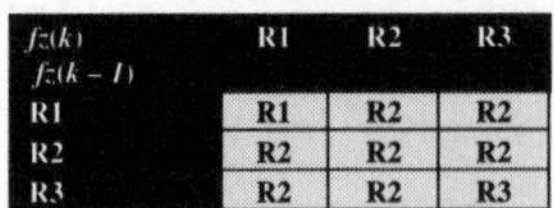

$fz(k)$ / $fz(k-1)$	R1	R2	R3
R1	R1	R2	R2
R2	R2	R2	R2
R3	R2	R2	R3

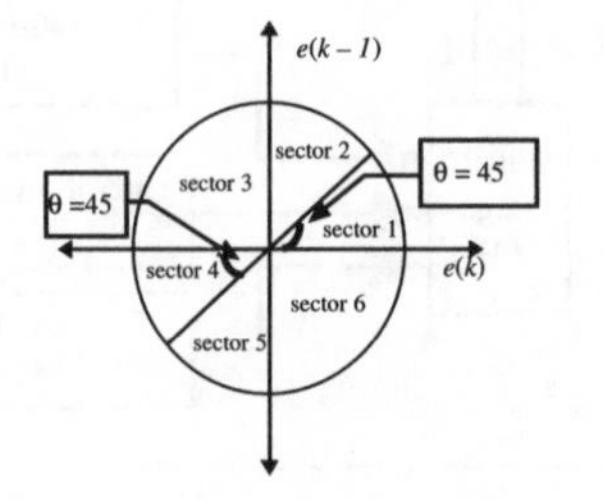

Fig. 5.11 *Operational sectors*

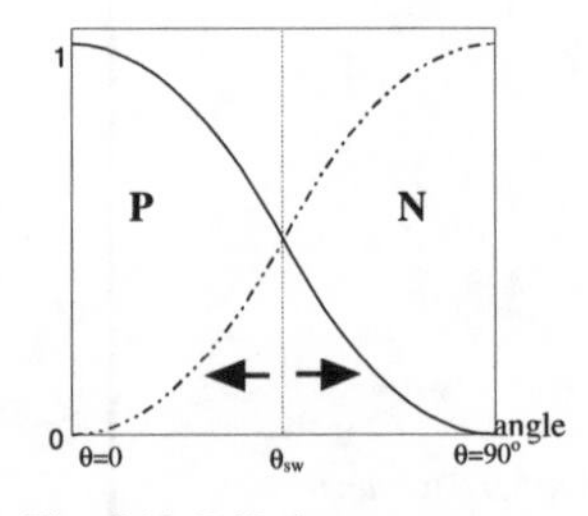

Fig. 5.12 *P-N classes*

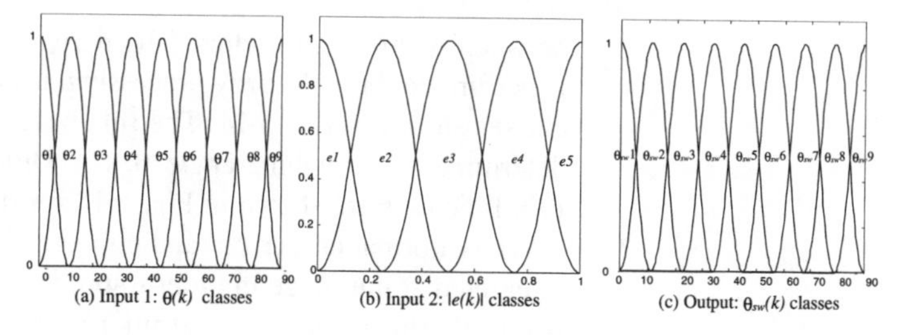

Fig. 5.13 *Increment–decrement classes*

Block 2 in Fig. 5.9 analyses the current sector of the fuzzy controller operation [Ghan95-2]. A definition of the sector is performed by considering two successive samples of the performance monitor ($e(t)$ and $e(t-1)$) and the acceleration (Δe) of this monitor ($\Delta e(t) = e(t) - e(t-1)$). The division of the error signal plane into sectors is presented in Fig. 5.11.

Block 3 in Fig. 5.9 is a decrement/increment block. The objective of this block is to produce a signal that will increase or decrease the input and/or output ranges. The tuning signal value generated by this block depends on how much a system error signal is changed during the current sampling period. The value of change is characterised by two parameters:

- $\theta(k)$, the angle between the two consequent positions of the error signal on the operational plane (Fig. 5.11);
- $|e(k)|$, the absolute value of the error signal.

Table 5.3 Rule base for deriving θ_{sw}

Decrement						Increment					
$\theta(t)$ \ $\lvert e(t)\rvert$	e_1	e_2	e_3	e_4	e_5	$\theta(t)$ \ $\lvert e(t)\rvert$	e_1	e_2	e_3	e_4	e_5
θ1	θ_{sw1}	θ_{sw2}	θ_{sw2}	θ_{sw3}	θ_{sw3}	θ1	θ_{sw1}	θ_{sw2}	θ_{sw2}	θ_{sw3}	θ_{sw3}
θ2	θ_{sw2}	θ_{sw3}	θ_{sw3}	θ_{sw4}	θ_{sw4}	θ2	θ_{sw2}	θ_{sw3}	θ_{sw3}	θ_{sw4}	θ_{sw4}
θ3	θ_{sw3}	θ_{sw4}	θ_{sw4}	θ_{sw4}	θ_{sw4}	θ3	θ_{sw3}	θ_{sw4}	θ_{sw4}	θ_{sw5}	θ_{sw5}
θ4	θ_{sw4}	θ_{sw5}	θ_{sw5}	θ_{sw5}	θ_{sw5}	θ4	θ_{sw4}	θ_{sw5}	θ_{sw5}	θ_{sw5}	θ_{sw5}
θ5	θ_{sw5}	θ_{sw5}	θ_{sw5}	θ_{sw5}	θ_{sw5}	θ5	θ_{sw5}	θ_{sw5}	θ_{sw5}	θ_{sw5}	θ_{sw5}
θ6	θ_{sw6}	θ_{sw5}	θ_{sw5}	θ_{sw5}	θ_{sw5}	θ6	θ_{sw6}	θ_{sw6}	θ_{sw5}	θ_{sw5}	θ_{sw4}
θ7	θ_{sw7}	θ_{sw6}	θ_{sw6}	θ_{sw6}	θ_{sw6}	θ7	θ_{sw7}	θ_{sw6}	θ_{sw6}	θ_{sw6}	θ_{sw6}
θ8	θ_{sw8}	θ_{sw7}	θ_{sw7}	θ_{sw6}	θ_{sw6}	θ8	θ_{sw8}	θ_{sw7}	θ_{sw7}	θ_{sw6}	θ_{sw6}
θ9	θ_{sw9}	θ_{sw8}	θ_{sw8}	θ_{sw7}	θ_{sw7}	θ9	θ_{sw9}	θ_{sw9}	θ_{sw9}	θ_{sw9}	θ_{sw9}

Two nonlinear classes (P-N) (Fig. 5.12) are used to derive the decrement/increment values [Ghan95-2] and $\theta(k)$ is scaled between 0–90° in all sectors. The derivation of the tuning signal is performed through fuzzy processing, using the input and output classes shown in Fig. 5.14. The rule base of Table 5.3 is used to determine the switching angle $\theta_{sw}(k)$. Different arrangements of P-N functions are shown in Fig. 5.15 for different θ_{sw}.

The information derived and collected from previous stages is integrated at this stage in order to produce the final tuning signal for both the input and output ranges. Processing the full information is summarised in the logic flow diagram of Fig. 5.15. The processing at this stage relies on two main factors, the reference signal and the performance monitors. The logic flow diagram in Fig. 5.15 demonstrates the final stage processing, used in the testing examples, where a reference signal of positive and negative values is applied and the performance monitor is the error signal (e), where e is a difference between the reference and the main controller output signals.

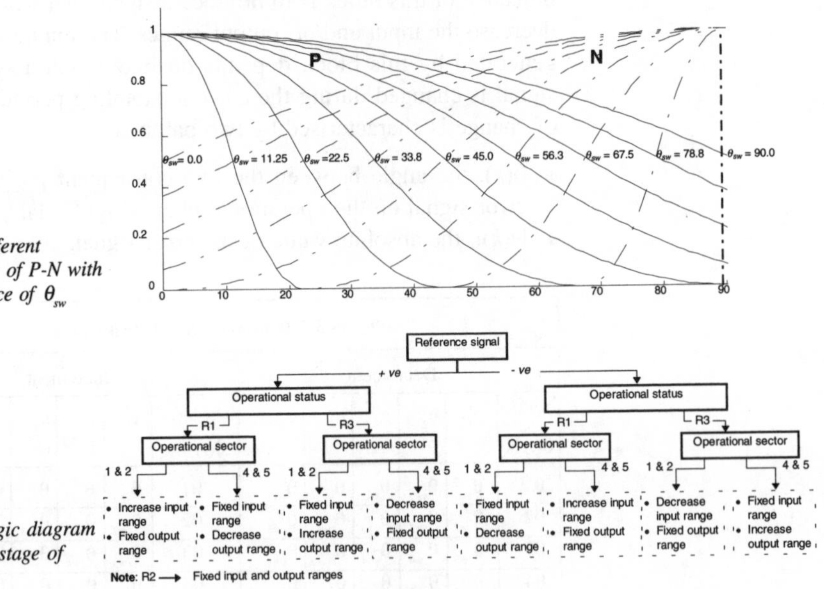

Fig. 5.14 *Different arrangements of P-N with different choice of* θ_{sw}

Fig. 5.15 *Logic diagram for the final stage of tuning*

5.3 Artificial neural networks and neuro-fuzzy controllers

5.3.1 What is a neural network?

There are many answers to this question. Many definitions of neural networks assume that you already know the basics of such networks, or that you are at least mathematically literate and comfortable with calculus and differential equations. Although mathematical models of computation are valuable, it is easy for people who are not specialists to be intimidated by them. Other definitions tend to be jargon, and still others end up being esoteric or vague. Often they talk about a new computing architecture inspired by biological models. So following [Nel91] let me give you a few definitions, describing different aspects of the nets.

The artificial neural network (ANN) is:

- A new form of computing, inspired by biological models.
- A mathematical model, composed of a large number of processing elements organised into layers.
- A computing system, made up of a number of simple, highly interconnected processing elements, which processes information by its dynamic state response to external inputs.

The first definition tends to consider the origin of the ANN. The second one considers ANN as a mathematical model of the object. The third one refers to a structure and a technological implementation.

I guess the name was inherited from its biological prototype.

Correct. Most of the textbooks on this subject include a description of a biological neuron and consider a human brain as a network of such neurones. The authors assume that a biological neuron is so simple and well known to a reader, that it helps to explain such a complicated technological phenomenon as a ANN. To differentiate the technology from nature, the corresponding computing system is called an artificial neural network. The ANN is supposed to consist of artificial neurones or processing elements (PE). These artificial neurones bear only a modest resemblance to the real things. Processing elements model approximately three of the processes, out of the 150 we know, that neurones perform in the human brain.

5.3.2 ANN structure

The PE handles several basic functions:

- It evaluates the input signals, determining the strength of each one.
- It calculates a total for the combined input signals and compares that total against some threshold level.
- Depending on the result, it determines what the output should be.

Any PE has many inputs and one output (see Fig. 5.16) All of them should come into our PE simultaneously. As a response to the input varies, a neuron either 'fires' or 'doesn't fire', depending on some threshold level.

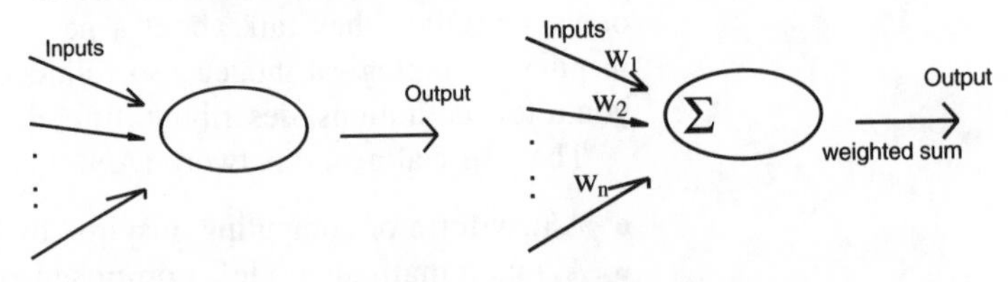

Fig. 5.16 *Processing element (PE)*

Fig. 5. 17 *Processing element with an output determined as a weighted sum of the inputs*

How does a PE calculate the output value using the input values?

Let us consider the simplest case. Here the output is just a weighted sum of the inputs (see Fig. 5.17). If we denote the ith input as x_i and the output as y, then we can write the mapping from the inputs to the output performed by the PE in this case as:

$$y = f(x_1, x_2, \ldots, x_n) = w_1 \times x_1 + w_2 \times x_2 + \ldots + w_n \times x_n$$

You see that this is a linear function, and the linear ANN can be constructed in this way. You may have another function, not a sum but a product or a logical function, for example.

Which function will we have then?

A nonlinear one and a nonlinear network. Another way to create a nonlinear network is to compare the mapping result with the threshold (Fig. 5.18). As a result of a comparison, a special impulse (an excitation signal) is generated if the weighted sum of the inputs is higher than the threshold value (the neuron is fired). I should mention that the threshold usually is chosen as a function of the inputs and in control applications is called the transfer function.

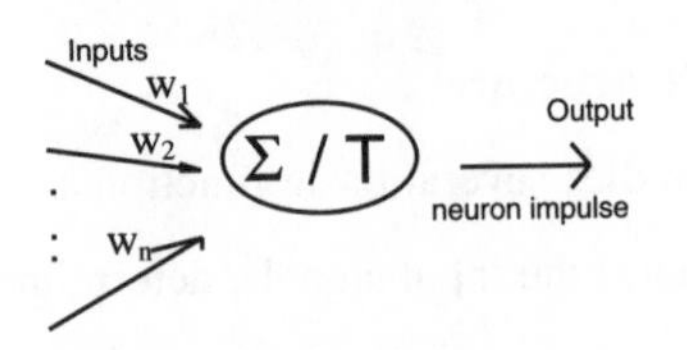

Fig. 5.18 *Processing element with an output determined as a comparison result of the weighted sum of the inputs with the threshold*

How do we determine this function?

This is a good question as it represents a current research area. The choice of a linear function is pretty boring, but it will allow you to model just linear systems.

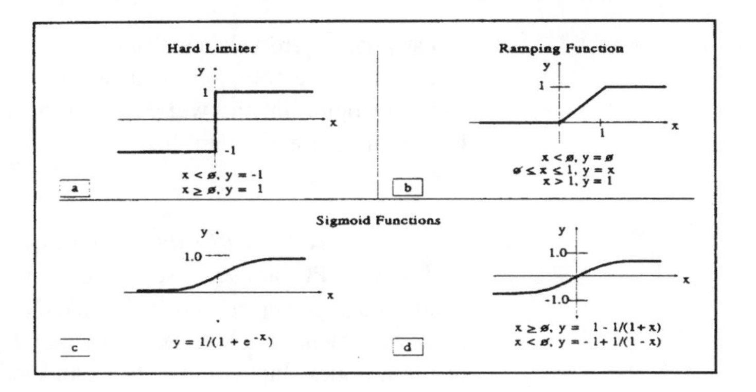

Fig. 5.19 *Different ways to determine the threshold (transfer) function*

Why not choose just the threshold function as in Fig. 5.19a?

We may have a lot of problems with this function because we are not able to take the derivative from it near the transformation point. Learning methods based on a gradient descent include differentiating. That's why some more complicated functions given on Figs 5.20c and d are applied in modern ANNs.

The transfer function could be something as simple as 0. The output then depends upon whether the result of the summation is positive or negative. The network could output 1 and –1, or 1 and 0, etc., and the transfer function would then be a 'hard limiter' or 'step' function (Fig. 5.19a). It has only a binary output. Another type of transfer function is the threshold or ramping function (Fig. 5.19b), which could mirror the input within a given range, say from zero to one, but could operate as a hard limiter outside that range. It is a linear function that has been clipped to the minimum and maximum values, which then makes it nonlinear. Yet another option would be a sigmoid or S-shaped curve (such as shown in Figs 5.19c and d). The curve approaches minimum and maximum values at the asymptotes.

That sounds pretty interesting but does not explain how an ANN can be used for our purposes.

New effects appear when you connect PEs into a network and if we attach some local memory to our PE, we can store results of previous computations and modify the weights used as we go along. This ability to change the weights allows a PE to modify its behaviour in response to its inputs, or to 'learn'. For example, suppose a network identifies a dog as 'a cat'. On successive iterations, connection weights that respond correctly to dog images are strengthened, those that respond to other images, such as cat images, are weakened until they fall below the threshold level. It is more complicated than just changing the weights for a dog recognition; the weights have to be adjusted so that all images are correctly identified.

How can we connect PEs into a network?

Any way. PEs are supposed to compose a layer, and layers are connected together composing an ANN. Suppose we take one processing element and combine it with other PEs to make a layer of these nodes. Inputs could be connected to many nodes with various weights, resulting in a series of outputs, one per node.

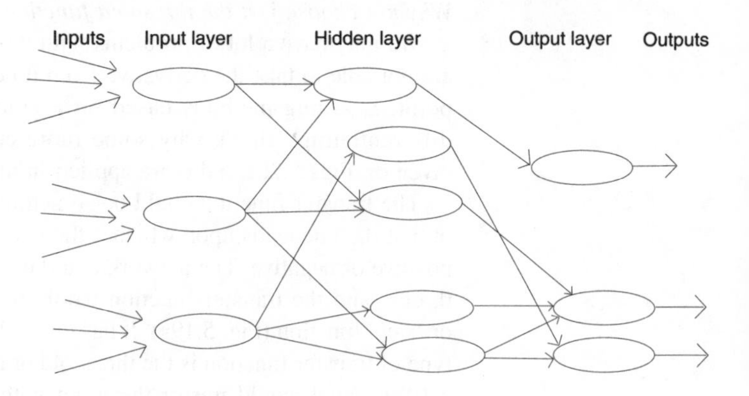

Fig. 5.20 *ANN structure*

Since each PE is a non-linear block, such an arrangement of PEs results in a highly non-linear network.

Carrying the design yet another step, we can interconnect several layers. The layer that receives the inputs is called the input layer. It typically performs no function other than buffering of the input signal. The network outputs are generated from the *output layer.* Any other layers are called hidden layers because they are internal to the network and have no direct contact with the external environment. Sometimes they are likened to a 'black box' within a network system. But just because they are not immediately visible doesn't mean you cannot examine what goes on in those layers. There may be several hidden layers.

The connections are multiplied by the weights associated with that particular node with which they interconnect. They convey analog values. Note that there are many more connections than nodes. The network is said to be fully connected if every output from one layer is passed along to every node in the next layer.

5.3.3 ANN types

Can the nodes be connected in any way? Is any order necessary?

It depends on the type of the particular network. Actually a lot of different ANN have been proposed. Some of them are given in Table 5.4. Sometimes it is very hard to answer even a simple question. Anyway we will try to give just a very general classification. On the basis of the connection types ANN can be divided into feedforward and recurrent. If the output of each node propagates from the input side (left) to the output side (right) unanimously then the ANN is called feedforward (Fig. 5.21). If there is any link between nodes directed from the right to the left, this ANN has a recurrent type (Fig. 5.22).

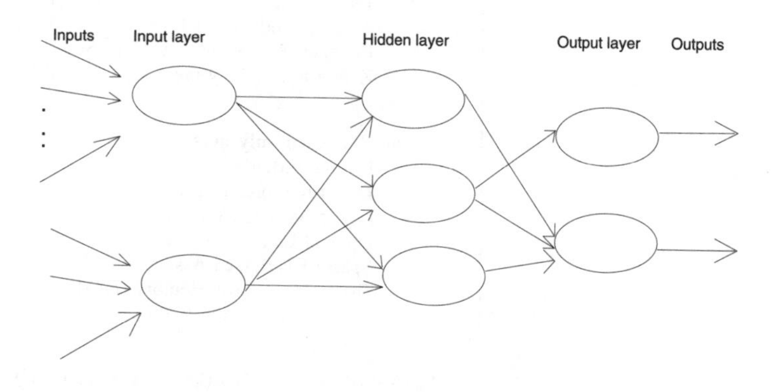

Fig. 5.21 *A feedforward network*

The life of any ANN includes two phases: learning and application. Before one starts applying an ANN, it should be trained. The training procedures are determined by a learning regime and can be quite different from each other.

Generally, the goal of a neural network is to modify the weights to minimise the response error. Nudge the weights in the direction that will minimise the difference between the measured and desired outputs. The technique for modifying weights in the appropriate direction and magnitude (called a credit assignment problem) eluded researchers for years. However, the recent development of the back propagation learning scheme has solved this problem. Since the advent of 'back-prop', neural networks have enjoyed renewed emphasis and application.

Table 5.4

UNSUPERVISED LEARNING **(i.e., without a 'teacher')**

Feedback nets
- Additive Grossberg (AG)
- Shunting Grossberg (SG)
- Binary Adaptive Resonance Theory (ART1)
- Analog Adaptive Resonance Theory (ART2, ART2a)
- Discrete Hopfield (DH)

UNSUPERVISED LEARNING **(i.e., without a 'teacher')**

Feedback nets
- Continuous Hopfield (CH)
- Discrete Bidirectional Associative Memory (BAM)
- Temporal Associative Memory (TAM)
- Adaptive Bidirectional Associative Memory (ABAM)
- Kohonen Self-organising Map (SOM)
- Kohonen Topology-preserving Map (TPM)

Feedforward-only nets
- Learning Matrix (LM)
- Driver-reinforcement Learning (DR)
- Linear Associative Memory (LAM)
- Optimal Linear Associative Memory (OLAM)
- Sparse Distributed Associative Memory (SDM)
- Fuzzy Associative Memory (FAM)
- Counterpropagation (CPN)

SUPERVISED LEARNING **(i.e., with a 'teacher')**

Feedback nets
- Brain-State-in-a-Box (BSB)
- Fuzzy Cognitive Map (FCM)
- Boltzmann Machine (BM)
- Mean Field Annealing (MFT)
- Recurrent Cascade Correlation (RCC)
- Learning Vector Quantisation (LVQ)

Feedforward-only nets
- Perceptron
- Adaline, Madaline
- Backpropagation (BP)
- Cauchy Machine (CM)
- Adaptive Heuristic Critic (AHC)
- Time Delay Neural Network (TDNN)
- Associative Reward Penalty (ARP)
- Avalanche Matched Filter (AMF)
- Backpercolation (Perc)
- Artmap
- Adaptive Logic Network (ALN)
- Cascade Correlation (CasCor)

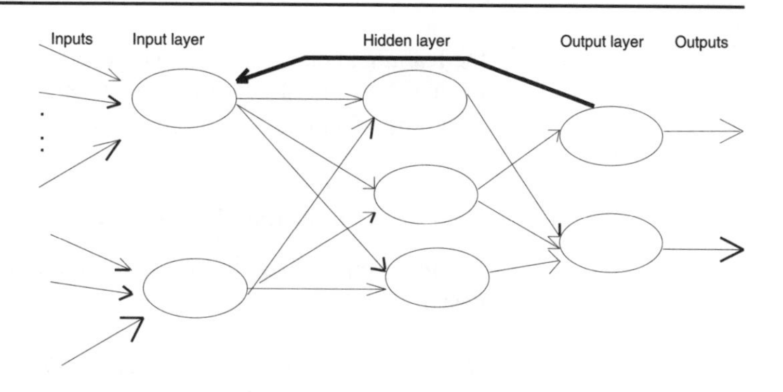

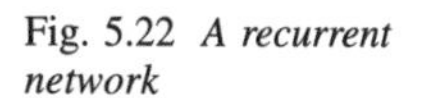
Fig. 5.22 *A recurrent network*

How does this back propagation work?

The method is based on an analysis of how a change in any particular weight influences the output of the network. After such analysis is done, the designer understands how to change the weights to achieve the specified values for the outputs.

What does such analysis include?

First of all, one must construct a chain similar to Fig. 5.23. This chain examines the influence of any weight factor on the output value and, hence, on the error value. We consider as an error the difference between the actual output and the desired one.

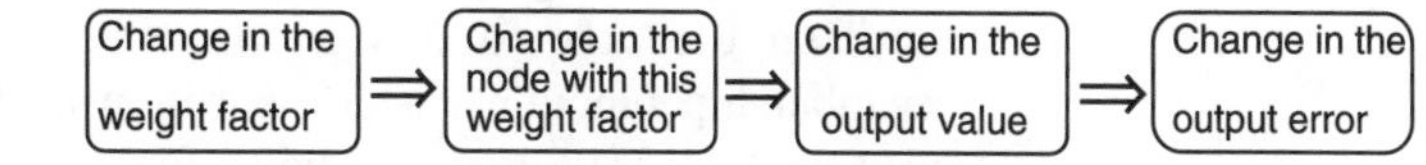

Fig. 5.23 *ANN analysis algorithm*

Is this just the way for forward propagation?

Yes it is. The way backward will include a definition of how any particular weight factor should be changed to minimise the error. Here one must go through these chains in a reverse order. The so-called ordered derivatives originating from ordinary partial derivatives, form a base of the process.

This process is a learning process, of which there are two types: supervised and unsupervised. In a supervised mode, the actual output of the ANN is compared with the desired output. The ANN must be trained before it starts being used. In an unsupervised mode, there are no external factors influencing weight adjustment.

I still do not completely understand.

Neither do I. At present, this learning mode is not well understood and is still a research subject. Generally, neural networks can be considered as trainable dynamic systems where the learning, noise tolerance and generalisation abilities grow out of their network structures, their dynamics and their distributed data representation.

5.3.4 ANN application in fuzzy controller design

We must come back to ANN application in fuzzy controller design.

How are ANNs applied in fuzzy controller design?

Once again by different ways. According to [Tak95] there are three main approaches:

- fuzzy systems where ANN learn the shape of the surface of membership functions, the rules and output membership values;
- fuzzy systems that are expressed in the form of ANN and are designed using a learning capability of the ANN;
- fuzzy systems with ANN which are used to tune the parameters of the fuzzy controller as a design tool but not as a component of the final fuzzy system.

In the first two approaches, ANNs become a component of the whole neuro-fuzzy system. In the first approach, an ANN is applied directly to design nonlinear multidimensional membership functions, which partition an input space. In this way, an ANN is utilised to conduct fuzzy reasoning [Tak91]. The rather famous ANFIS systems [Jang95] are an example of the second approach.

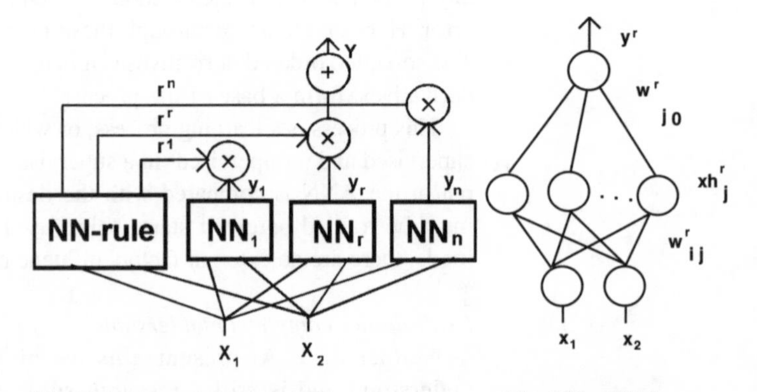

Fig. 5.24 *A neuro-fuzzy controller*

Fig. 5.25 *ANN for one rule implementing*

5.3.5 ANFIS architecture

Let us consider the example of a Sugeno fuzzy model given in [Jang95]. Here the ANN is used to realise fuzzy processing of typical fuzzy rules like

If x_1 is A_1 **and** x_2 is A_2 **then** $y_1 = w_1 \times x_1 + w_2 \times x_2 + r_1$
If x_1 is B_1 **and** x_2 is B_2 **then** $y_2 = v_1 \times x_1 + v_2 \times x_2 + r_2$.

To realise this processing, the ANN similar to Fig. 5.26 has been developed.

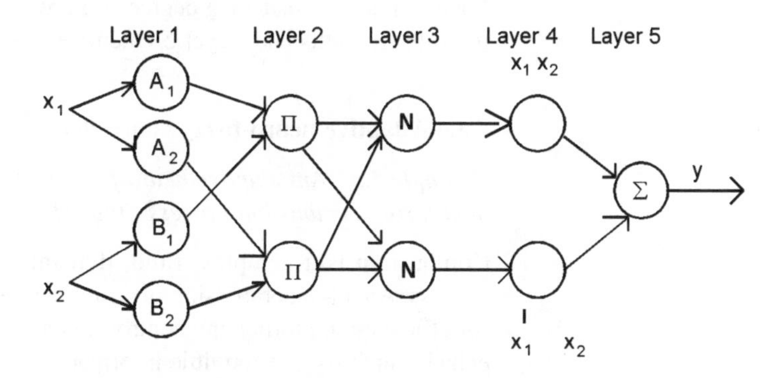

Fig. 5.26 *ANFIS architecture*

The first layer is the membership layer. An output of any node in this layer gives the membership degree of an input. The second layer is the multiplication layer. Every node here multiplies the inputs or membership degrees and produces the firing strength of the rule or the degree in which the corresponding rule is fired.

Why is multiplication applied? I understand that any t norm operation can be used.

You are absolutely right. Multiplication has been chosen in this application to represent the *t* norm operation. The third layer is a normalising layer. It calculates a ratio of the particular rule-firing degree to the sum of all rule degrees. The fourth layer is the output calculating one. Any node here is an adaptive node with the output calculated according to the formula:

$$f = V\,(w_1 \times x_1 + w_2 \times x_2 + r_1)$$

where V is the output of the third layer node and x_1 and x_2 are the inputs of the network.

Where does this polynomial come from?

This is the consequent part of the Sugeno fuzzy system which we want to model. Finally, the only node of the fifth layer calculates the overall output as the sum of all incoming signals. So the ANN has been constructed that operates exactly (functionally) as the Sugeno system.

There are different ways of creating a neuro-fuzzy system. One is the ANN-fuzzy controller structure proposed in [Lin94]. This structure also has five layers but these are a little bit different. Layer 1 is a transparent layer. It just transmits input values directly to the next level. Layer 2 calculates the membership degrees, Layer 3 finds the matching degrees for any rule, Layer 4 integrates the rule strengths and Layer 5 calculates the output.

5.3.6 Adaptive neuro-fuzzy controller

Example 5.3 *An adaptive neuro-fuzzy system. Fuzzy control for monitoring combustion process [Tao94]*

Combustion is a complex, fluid, dynamic, reactive multiphase process. An all-encompassing measurement technique is the basis for efficient monitoring and control. Local point measurements or exhaust analysis give too little information about the process state. Therefore, the entire visual information for process control needs to be applied. Image processing methods are used for the recognition of non-optimal burning states. The neuro-fuzzy controller has to learn and achieve an optimal burning state which is set up by a human expert. The optimal and the current states are determined with a small number of characteristics which are to be measured. The differences between them are applied as controller inputs. Implementation of the fuzzy controller (Fig. 5.27) in the form of a neural network provides the possibility that the system can learn from the environment and improve the robustness against environmental disturbances through updating network weights.

The system determines a control action by using a neural network which implements a fuzzy inference. In this way, the prior expert's knowledge can be incorporated easily. The controller has two states, a learning state and a controlling state. In the learning state, the performance evaluation is carried out according to the feedback which represents the process state. The original model was proposed in [Lin91, Ber92]. If input–output training data (a teacher) is available, the performance can be assessed easily, and supervised learning can be employed.

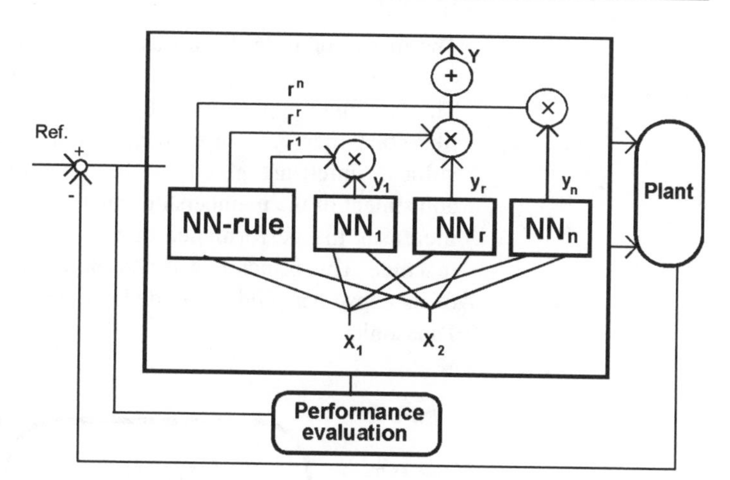

Fig. 5.27 *Neuro-fuzzy controller for combustion monitoring*

Otherwise, the performance evaluation only gives a reinforcement signal where reinforcement learning can be applied. In the controlling state, the controller simply does fuzzy inference without adjusting its parameters. The neural network consists of a few layers. Nodes in Layer 1 are input nodes, representing input linguistic variables. They transmit sensor readings to the next layer which performs fuzzification. Nodes in Layer 3 are rule nodes and ones in Layer 4 perform defuzzification.

5.3.7 Application examples

Example 5.4 *Neuro-fuzzy controller for a washing machine [Tak95-2] (Matsushita Electric Group)*

Fig. 5.28 *ÖKO LAVAMAT 6955 sensor logic washing machine for stains*

This model needs less water, energy and detergents, yet gets the wash spotlessly clean.

Its features:

- New 'stains program' for effective, specific stain removal.
- Consumption: 50 litres of water and 1.7 kWh.
- Sensor-controlled automatic dosing by fuzzy logic.

The fuzzy controller for a washing machine applies three inputs: clothes mass, impurity of water and time differential of impurity. The last input application allows a construction of a fuzzy PD-like controller. Any input description includes three linguistic values, which are given by their membership functions. An adjustment of the membership functions, performed by the ANN, increases the performance of the controller and the washing machine. This principle was chosen by Matsushita for the design of its consumer products traded overseas under the trademark of Panasonic.

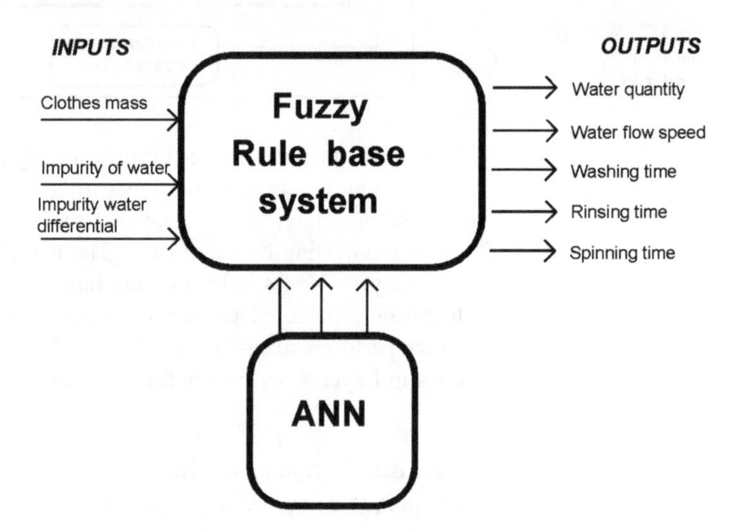

Fig. 5.29 *Neuro-fuzzy controller for a washing machine (Matsushita Electric Group)*

Example 5.5 *A neuro-fuzzy controller for a washing machine produced by Hitachi [Tak95-2]*

Hitachi has chosen another way to design a neuro-fuzzy controller. It applies the ANN to correct the fuzzy controller output values. This way means the company does not need to redevelop a whole fuzzy controller. It is particularly important in the case when the input set is changed, for example, because a new sensor has been added. Due to a strong competition in the consumer products market, the designers have to modernise products very often. In this design, just the more flexible ANN part should be modernised.

Here, the sensor that measures the water temperature has been replaced with a sensor for air temperature. However, the old fuzzy controller continues to be used. The ANN part is applied to modify the output values (see Fig. 5.30).

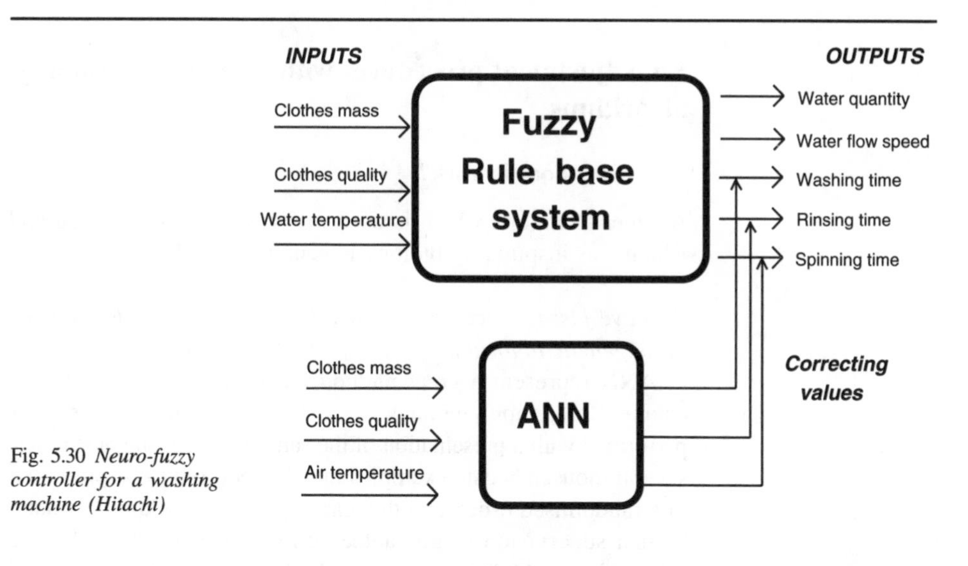

Fig. 5.30 *Neuro-fuzzy controller for a washing machine (Hitachi)*

I understand that fuzzy logic and ANN work simultaneously or in parallel. It saves a lot of time.

Right. It did in these examples but generally it doesn't have to.

Example 5.6 *Sanyo rotating electric fan.*

This fan should rotate itself towards the user. The user's position is identified by a remote control signal origin. The fan has three infrared sensors where the signals are used to calculate a distance to the user using a fuzzy system. Then this distance and the ratios of sensor outputs are used by an ANN to compute the fan direction (Fig. 5.31). This combination has allowed significant improvement in the quality of a distance calculation.

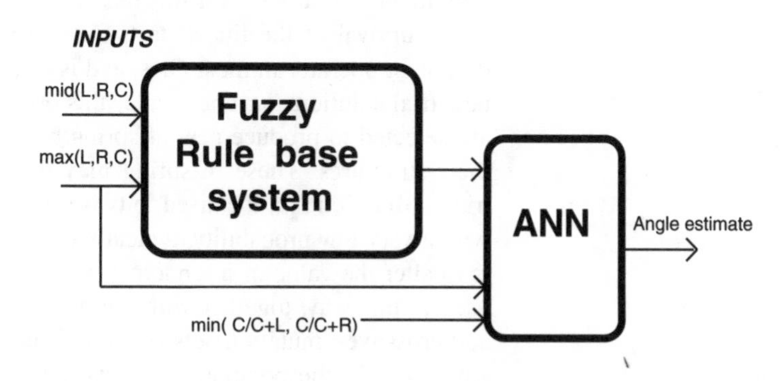

Fig. 5.31 *A neuro-fuzzy controller for a rotating fan*

5.4 Adjustment procedures with genetic/evolutionary algorithms

5.4.1 How does it work?

Another technique which is used in fuzzy controller design and which was inspired by biology is genetic algorithms.

We have just finished with artificial neurones. What do we need these genetic algorithms for? Isn't ANN enough?

ANN represent a great method but it has its limitations, of course. Before applying, any ANN should be trained. Training is performed with a presentation of the sampling data. In some cases, several thousands data examples may be required to be presented in a randomised order, and the learning techniques often demand human supervision to guarantee convergence [Link95]. So the application of ANN may require a high computational power and a long period for training, which are not available in some control applications. There is a need to optimise the training process, and genetic algorithms (GA) application is one of these ways.

Explain about genetic algorithms.

The GA method imitates natural evolution processes, and hence includes operations such as reproduction, crossover and mutation. A conventional GA has four main features: population size, reproduction, crossover and mutation. Reproduction is a process in which a new generation of population is formed by the stochastic selection of individuals from an existing population, based on their fitness.

This process results in individuals with higher fitness values being more prevalent in the next generation, while low fitness individuals are less so; for this reason it is sometimes referred to as a 'survival of the fittest' test. Crossover is perhaps the most dominant operator in most GAs, and is responsible for producing new trial solutions on-line. Under this operation, two individuals are selected to produce new offspring by exchanging portions of their structures. These offspring may then replace the existing trials. Mutation is a localised, bitwise operator, which is applied with a very low probability, typically 0.001 per bit or less. Its role is to alter the value of a random position in a gene-string. When used in this way, together with the other operators, reproduction and crossover, mutation acts as an insurance against total loss of any genes in the population, by its ability to introduce a gene which may not, initially, have existed or was lost in the course of application of other operators.

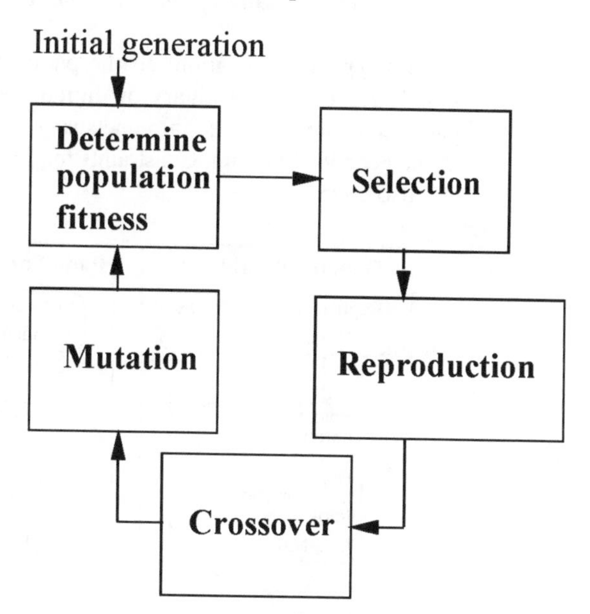

Fig. 5.32 *Genetic algorithm procedure*

For a given population of trials and set of operators, together with the evaluation function, a typical GA proceeds as follows:

- an initial random population of trials, $\Pi(0) = A_n(0)$, $n = 1,\ldots,N$, where N is the population size, is generated;
- for successive sample instances t, the performance of each trial, $\mu\ (A_n(t)), t = 0,1,2,\ldots$, is evaluated and stored;
- one or more trials are selected to produce offspring by taking a sample of $\Pi(t)$ using the probability distribution: $\rho\ (A_n(t)) = \mu\ (A_n(t))/\Sigma\ \mu\ (A_i(t))$
- one or more of the genetic operators are applied to the selected trials in order to produce new offspring, $A^\circ_m\ (T)$, $m = 1,\ldots, M$, where M is the number of offspring, which is usually equal to the number of selected parent trials;
- the next generation of population, $\Pi(t + 1)$ is formed by selecting $A_i(t) \in \Pi(t)$, $i = 1, \ldots, M$ to be replaced by the offspring, $A_j(t)$, $j = 1 - M$. The criterion for selecting which trials should be replaced may be random, on the basis of the least fit or some other fitness basis;
- the GA process is terminated after a pre-specified number of generations or on the basis of a criterion which determines convergence of the population.

5.4.2 GA and EA application in fuzzy controller design

How are GA actually applied for tuning the parameters of the fuzzy controller?

In a typical application, all the parameters which are to be tuned are coded as a binary or digital code. The codes of all parameters are combined into one string which is a chromosome of the genetic algorithm. Constraints for the controller parameters are also coded.

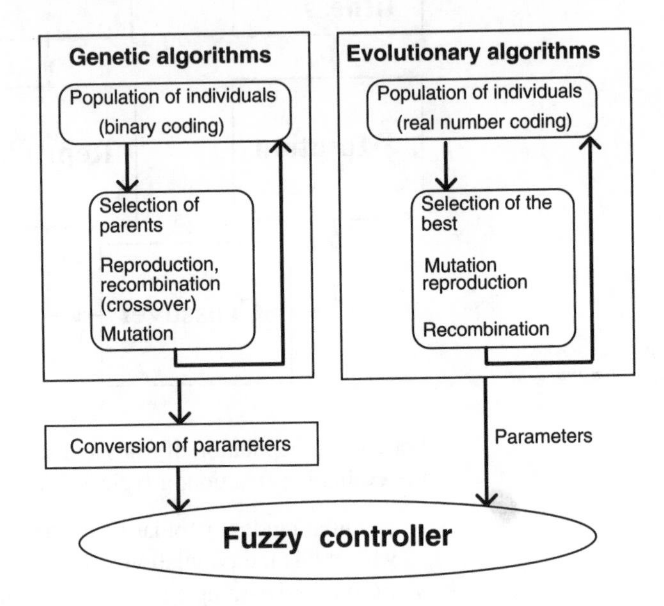

Fig. 5.33 *Structure of GA and EA*

What is the difference between genetic and evolutionary algorithms (EA)?

The strategies are pretty similar to each other. The main difference is a coding format of the parameters to be adjusted [Pres95]. GA suppose binary coding of the parameter string and EA apply real number coding.

Is that all? Is it so important?

Yes, it is, because this difference determines the application areas for each of them. EA, due to their coding, are better suited for searching the optimum on flat and continuous spaces, while GA have no problems with discontinuity. It may have consequences for a fuzzy controller parameter adjustment. EA are superior when just small changes are supposed, while GA should be used when selecting rules or membership functions initially.

How do these algorithms work in fuzzy controller design?

Let us consider the process in greater detail following [Huang95]. GA have been widely applied for the membership function adjustment. One can adjust the shape of membership functions or parameters of the shape selected. For example, if the triangular function has been chosen, the set of three parameters has to be evaluated. As the parameters of not one but a few membership functions are to be determined, all of them are concatenated as a string. This string is then encoded. The number of bits (or digits) in a code string determines the length of a chromosome – the parameter of GA.

Let us consider the operation with GA in greater detail, step by step (see Fig. 5.32). On the first step, a random number generator is applied to generate individuals – chromosomes. On the second step, the fitness of each chromosome is evaluated. The predetermined fitness function is used here to calculate how good the chromosome is. Criteria traditionally applied in control engineering can be utilised here as a fitness function, for example, the error signal. One has to remember, however, that the higher value of the fitness function should correspond to the better chromosome. So the error signal can be applied directly as a fitness function but should be converted. In the case of the error the fitness function could be chosen as e^{-ke} where k is a positive number and e is the error value.

At the next step the process of a parent selection is conducted. This process is similar to a spin of a roulette wheel. The selection algorithm is implemented based on the computed fitness of individuals. A new population pool is formed by several spins of a roulette wheel. Since the slot size is made to be proportional to the individual's fitness, most of the selected parents come to the new population pool with a higher fitness. Although the average fitness in the new pool is higher than in the original pool, some individuals may have the lower fitness. A pair of parents are then selected from this new pool for a further genetic operation.

The next step is the stage for a crossover and mutation process. A crossover point along the chromosome is randomly selected. A part of the genes on one side of this point is given to the child. The rest of the child's genes come from the same side of the other parent. Mutation plays a vital role in genetic optimisation as it introduces new genetic material that was lost by chance through poor selection of mates. However, one must be careful, as the overuse of mutation may distract a good chromosome in the next generation. The probability of mutation must be small enough to avoid degenerating the performance [Krist92]. So the mutation rate is usually selected to be about 0.0001–0.01. The number of individuals in each generation (the population size) is chosen on the base of the experiments. All the individuals in each generation have to be evaluated. One should understand that a large size may cause a long operation time. So usually the operation size is chosen between 50 and 500.

5.4.3 Application example

***Example** 5.7 Membership functions adjustment in aircraft flight control [Bour 96]*

Let us consider an adjustment of the membership functions for the fuzzy controller given in Example 3.5. The triangular membership functions are applied to describe input and output variables. There are two ways to achieve optimal membership functions:

- choose the entire membership functions as variables to be optimised;
- only choose the overlaps between membership functions as variables to be optimised.

We use the first method for the optimisation procedure. In this case, the triangle membership functions can have a variable width and be shifted along the x axis. This is the method of choosing all the points of the triangle shapes of the membership functions and moving these points along the x axis (as long as they do not exceed the upper and lower limits of the universe of discourse) until efficient membership functions are found. The triangles can change their variable base. Therefore, each triangle requires the definition of only one point to fix it. The entire set of fuzzy membership functions for a fuzzy controller must be represented as a binary string (of 0 and 1). To do this, the best choice is to use binary coding.

To apply GA, one needs initially to perform the following procedures

- Parameter coding: translate the fuzzy membership functions into strings.
- Initial generation: generate an initial population of 50 strings by using random numbers.
- Fitness function: decode each string into a set of actual parameters as membership functions and calculate the fitness value of each set by simulation.

The bit strings, representing the overlapping parameters, must then be judged and assigned a fitness value, which is a score representing a degree to which they accomplished the goal of defining high performance. However, the definition of the fitness function, that enables the GA to locate high performance and efficient membership functions, is application dependent. The performance index of the *i*th string (or the *i*th membership function) is defined as a trajectory length which is required to reach the goal point.

- Selection: copy strings into the mating pool according to their fitness value.
- Reproduction: copy strings with better fitness values into a new generation. We assume that the best 10 per cent of the generation of the size of 50 can be copied into the new generation.
- Generation of new strings: generate 90 per cent of 50 strings into a new generation.
- Crossover: select pairs of strings from the mating pool and exchange their corresponding portions of a string at a randomly selected position. The crossover probability of 0.8 is taken for simulations.
- Mutation: alternate the value of the randomly selected integer byte, this value may be changed into any number. The mutation probability of 0.1 is supposed.
- Iteration: repeat the process from the fitness function calculation to mutation until it reaches the predetermined ending condition.

6 FUZZY SYSTEM DESIGN SOFTWARE TOOLS

6.1 Fuzzy technology products classification

A fast-growing information technology industry has already developed and released about a dozen good design packages which are successfully used in different applications for fuzzy controller design. Among them are: RT/Fuzzy™ Toolbox for MATRIXx™ by Integrated Systems Inc., Fuzzy Logic Toolbox for MATLAB™ by The MathWorks Inc., FIDE™ by Aptronix, fuzzyTECH™ by Inform, a number of products by Togai InfraLogic Inc., Fuzzy Systems Engineering Inc., HyperLogic, etc. Some of them are specific for a fuzzy technology, others are universal and include a special fuzzy design toolbox. These products started appearing on the market in the late 1980s and early 1990s.

Table 6.1

Company	Products
Aptronix (USA)	FIDE
Bell Helicopter Textron (USA)	FULDEK
Byte Craft (Canada)	Fuzz-C
CICS Automation (Australia)	Fuzzy Control Toolbox for UNAC
Fuzzy Systems Engineering (USA)	Fuzzy Knowledge Builder, Fuzzy Decision Maker, and Fuzzy Thought Amplifier
FUZZYSOFT AG and GTS Trautzel GmbH (Germany)	FS-FUZZYSOFT
FuzyWare (USA)	FuziCalc
HIWARE (Switzerland)	HI-FLAG
HyperLogic (USA)	CubiCalc, CubiCalc RTC, CubiCalc RuleMaker, CubiCard, and CubiQuick
Inform (Germany)	fuzzyTECH
Integrated Systems (USA)	RT/Fuzzy in MATRIXx
Intelligent Machines (USA)	O INCA
KentRidge Instruments (Singapore)	FlexControl
Management Intelligenter Technologien (MIT) GmbH (Germany)	DataEngine
The MathWorks (USA)	Fuzzy Logic Tool Box for MatLab

Table 6.1 (continued)

Company	Products
MentaLogic Systems (Canada)	Quick-Fuzz, Auto-Fuzz, Multi-Fuzz, Flex-Fuzz, Process-Fuzz, Auto Sim-Fuzz, Fuzz-Drive, F-MTOS, CT-FLC
Metus Systems (USA)	FuzzySP
MODICO (USA)	FUZZLE
National Semiconductor (USA)	NeuFuz
Nicesoft (USA)	Decision Plus
Texas A&M University (USA)	FL Control
Togai InfraLogic (USA)	TILShell+ and other products
University of New Mexico (USA)	Fuzzy logic Code Generator (FLCG)

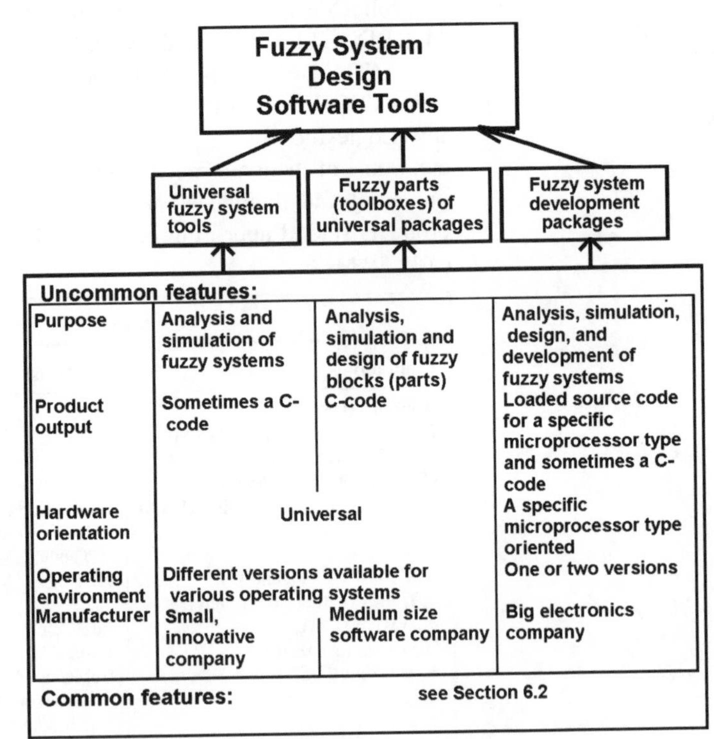

Fig. 6.1 *Classification of fuzzy system design tools*

The availability of the products on the market as well as their price means that we will not include in this text any homemade software tools for fuzzy controller design. Certainly, the main advice is to use one of these packages. The purpose of this chapter is to illustrate how. We will consider the features of the packages and give some examples of their applications. The main difference from other references available at the moment is that we try not to describe different packages but to point out some their common and specific features. This is the only way to preserve the value of this text for some time, at least. For more detail see the information booklets issued by companies, as well as an excellent survey in [Chiu95].

Table 6.1 contains the names of the some companies that produce fuzzy products. Please note that not all of these products and even companies are currently on the market. This table was compiled from different sources including [Chiu95, Jam94]. The trademarks given in the right-hand column of Table 6.1 are registered with the corresponding companies in the left-hand column. A little more detailed information about the companies including their addresses and some characteristics of their products is provided in Table 8.6.

Let us look at the different types of fuzzy products available nowadays (Fig. 6.1). Software tools for fuzzy system analysis (FuzzyCalc™, CubiCalc™, etc.) were historically the first generation of the products on the market. They were produced by small innovative companies. After that, companies producing general-purpose simulation and design tools (MATRIXx™, etc.) introduced fuzzy parts into their packages. One should note that despite the availability of universal fuzzy design software, many companies have chosen to invest money and time in the development of specialised, in-house fuzzy logic design tools. Companies such as Motorola, Intel, Siemens, SGS-Thomson, Texas Instruments and Hitachi have demanded customisation from the providers of commercial fuzzy tools because the off-the-shelf products were inadequate for their requirements. Similarly, National Semiconductor built its own in-house fuzzy software product to aid in the design and implementation of their semiconductor chips. This generation of commercial fuzzy logic design products had a limited aim: the design of simple control systems for specialised embedded microprocessors.

The list of the products and the review following in this chapter refer mainly to the universal design packages, that is, software tools which can be applied for different designs and which are not oriented towards use on a particular hardware platform. The most common features of the packages available are considered and illustrated with some examples in the following sections.

6.2 Main features of the fuzzy software tools

The goals of the design package are usually:

- speeding up a design process;
- making the design simple, user friendly and attainable;
- providing the possibility to apply fuzzy logic and control methods without a deep understanding of how they work.

Applications are often universal and can be applied in expert systems as well as control system design; only a few target control applications and control system design.

It is important to provide good service, pre- and after-sale customer support, and continuous improvement and development of the products.

The basic description and programming language is an object-oriented high level language that provides connections and links to other design tools, programming and simulation environment (e.g. FTL™ by Inform Software).

In order to achieve multisystem realisation, different versions are usually available for MS-Windows™, UNIX™ , and VME™ operating system environments.

A design package often includes other intelligent technology tools, first of all neural networks, and the option to integrate them with fuzzy design tools in order to improve and even optimise a design (e.g. Neural Networks toolbox in Matlab™, NeuroFuzzy™ module in fuzzyTECH™).

The package includes automation tools for all life phases of the product development, from design to implementation through simulation, debugging, optimisation for a target platform and verification. It supports a design environment where a designer does not have to write a single line of programming code.

The design package is almost always accompanied by the educational tools provided by the manufacturer. These tools can be used for learning the fundamentals of fuzzy logic and control, as well as for acquiring basic knowledge about using and applying this particular package. The tools can be provided as a written tutorial (Togai InfraLogic) or even as a computer educational kit (Fuzzy Logic Educational program by fuzzyTECH™ Explorer).

It supports an advanced man–machine graphical interface including all-graphical editors for different elements of fuzzy systems and other systems in all design phases. Also, it usually provides a user-friendly interface for any level user, sometimes with the possibility of producing an executable program code without any programming.

Different methods of fuzzy processing are supported, shapes of membership functions, fuzzification and defuzzification methods, etc., providing a user with many choices

The package is able to generate a portable C-code (almost all packages) and/or source codes oriented to different types of microprocessors, programmable controllers, etc. The type of code generated can be a main factor in choosing between the systems. They can interface with a large number of other general design tools such as MATLAB™, MS-Excel™, etc. Systems are often supplied in a package with the hardware products (different products have this option in various degrees from a board to a special control unit).

Many of the companies producing fuzzy design tools originate from university research and were founded by distinguished fuzzy logic and control academics and scholars (Zimmermann, Jamshidi, Goodwin, Welstead).

6.3 Realisation examples

The examples following illustrate different features of the design packages and give some more information about them.

Example 6.1 *Support of the entire development process including system analysis of the problem – FS-FUZZYSOFT design environment (Fig. 6.2)*

The first step in the design of a fuzzy-based system is an analysis of the problem. This results in a basic system outline structure, and includes its inputs and outputs. The specific knowledge of the system has to be described by means of linguistic variables and fuzzy rules. During an implementation of the project, several operators and parameters have to be defined, such as a fuzzy-specific inference method, or an internal resolution (to investigate system behaviour when implemented as a microcontroller with limited data resolution, e.g., only 8-bit data, no floating point hardware).

For the system test, simulation can be done by applying various signals, either in a closed loop, together with a model, or finally, on-line at the real process. An optimisation is done interactively, by use of the various sophisticated debugging tools. As the final step in the design of a fuzzy system, the system is implemented, either by generating the C-code or the code for a specific microcontroller.

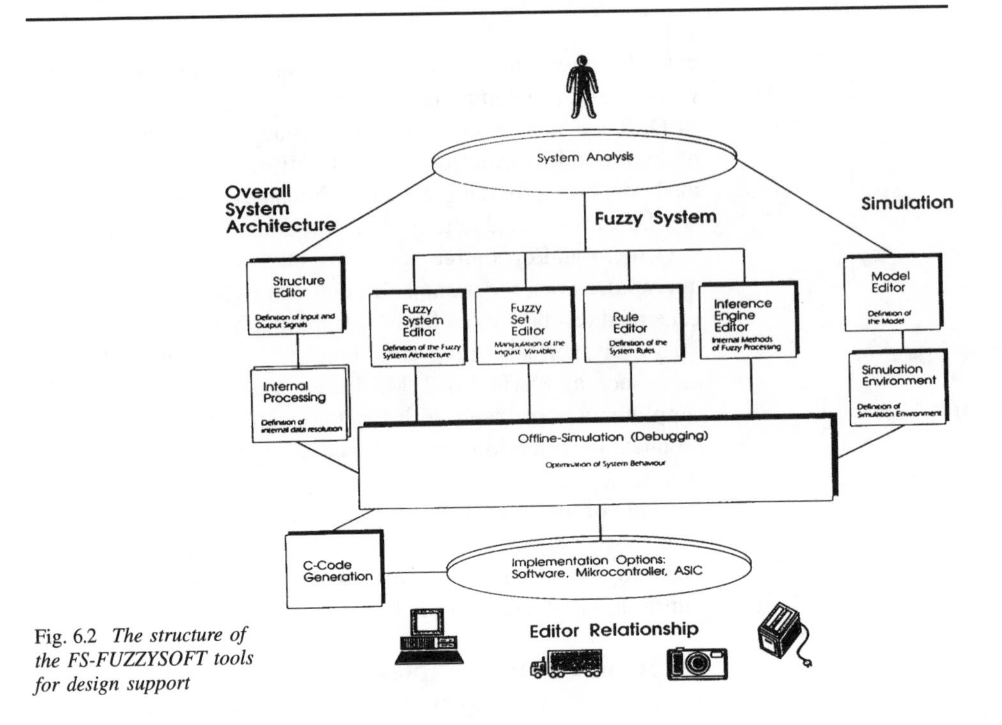

Fig. 6.2 *The structure of the FS-FUZZYSOFT tools for design support*

Example 6.2 *Incorporation of the fuzzy system block into modelling and simulation environment – RT/Fuzzy™ block in MATRIXx™*

The RT/Fuzzy block has been developed mainly for control applications. It is constructed as just one of the blocks among others to model and simulate a control system in the special environment called SystemBuild™. RT/Fuzzy has become an element of the library which includes linear dynamic systems, nonlinear elements, signal generators, mathematical functions and signal junctions. This is particularly convenient for rapid prototyping. The modularity principle is applied to the construction of the block itself as well. Any particular block can be accessed by clicking a corresponding icon. The rapid prototyping is highly recommended by the software authors as the method for applying fuzzy logic to describe parts of the whole system that are not yet fully understood. So it allows an approximation of the system behaviour with a fuzzy logic model.

The fuzzy block itself contains the rules and the data about the inputs and outputs. A user has an opportunity to enter different rules and data, change and edit them. User-definable functionality generally is one of the construction principles of these packages. To define the fuzzy system, a user has to fill in a set of special forms interconnected to each other.

After the design is completed, the fuzzy controller becomes just one of the blocks in the entire model. Examples of such models are seen in Figs. 4.1 and 5.6.

Example 6.3 *User education with specially provided teaching tools – Fuzzy Logic Computer educational kit by Aptronix.*

This kit accompanies the FIDE™ design package but is available separately. It presents, in an interactive mode, basic knowledge in fuzzy theory, fuzzy system analysis and design. It includes some computer-based tutorials covering fuzzy logic theoretical fundamentals and fuzzy technology. The kit contains information about main applications and requires the user to design simple systems. It also includes a number of self-assessment questions to test understanding.

The first part of the kit presents the mathematical fundamentals of the fuzzy logic and fuzzy sets theory. The presentation is built as a comparison between crisp (classical) sets and fuzzy sets, classical logic operations and fuzzy logic operations. This construction allows the user both to use his or her knowledge of classical mathematics in learning new theory and to refamiliarise the user with classical set and logic theory. Then come the chapters specifically for fuzzy logic. The first part serves as the theoretical introduction for solving application problems, particularly problems of fuzzy control.

The main part of the kit is occupied by fuzzy control problems. Because of the practical orientation, the main domain in this part is devoted to the design of fuzzy logic controllers. It incorporates the manuals on FIDE fuzzy products and their application in fuzzy controller design.

The kit runs under MS-Windows. The computer realisation provides:

- an interaction between the user and a computer;
- a graphical presentation of educational materials;
- a simple menu-driven control of the computer session, with an adaptation to the user's level.

Example 6.4 *High-level object oriented programming language for design and development, – the TIL Fuzzy Programming Language™ (FPL) by Togai InfraLogic*

The TIL FPL is specifically designed for the implementation of fuzzy knowledge bases. Programming consists of combining special language objects. There are nine special objects: project, package, source, fuzzy, var, connect, rule, fragment and member. As there are special hierarchical relationships between different objects (see Fig. 6.3), specific rules are applied, where an object can be defined.

The project object is the top level object. There should be only one project in the file, which contains information necessary to define a fuzzy system. The package object can be applied to divide a fuzzy system into subsystems and to create a hierarchical structure. This object is used to describe a part. Fuzzy objects define fuzzy rule bases and rules objects define rules. To embed source codes into rules bases, fragment objects are applied. Var objects correspond to the variables in conventional programming languages. Member objects are used to define membership functions.

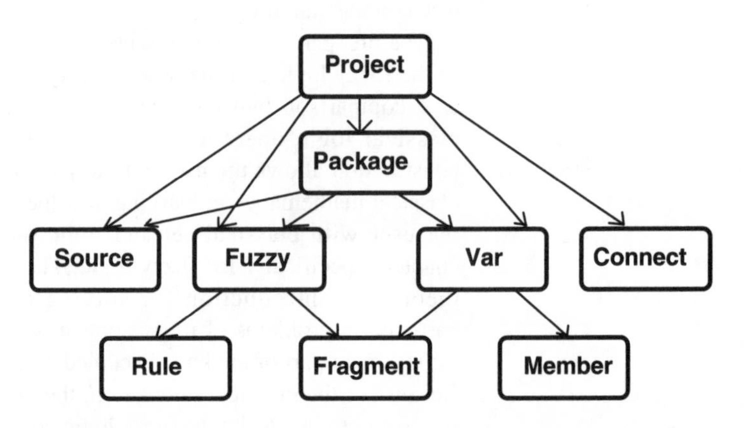

Fig. 6.3 *FPL object structure*

Example 6.5 *Inclusion of software tools for all life stages of the design: from the project description to implementation – FIDE™ by Aptronix*

FIDE (Fuzzy Inference Development Environment) integrates system design, tuning and simulation of a fuzzy system tools. FIDE tools work on two levels: FIDE inference unit (FIU) design and an overall system design.

To develop a FIU one has to:

- create and edit a FIU;
- debug, tune and simulate;
- generate a real-time code (for single FIU applications).

To design the whole system one may:

- graphically link FIUs and conventional execution units;
- debug and tune an overall system;
- make an overall system host executable code;
- run the overall system simulation while tuning FIUs;
- generate a real-time code for each FIU.

At the unit level, a single FIU is developed by specifying the membership functions and writing the rules. Optionally, one can develop a system for an application. The system level may consist of multiple inference units and conventional execution units. FIDE links previously compiled and tested FIUs and non-fuzzy units into a coherent whole design.

Example 6.6 *User-friendly design and advanced graphical interface – CubiCalc™*

CubiCalc provides a text editor for entering the rules as if–then statements and a graphic point-and-click editor for defining membership functions as connected line segments. Rule weights and linguistic hedges are supported. Rule weights can be entered by a user interactively or from other modules. While defining a membership function, the editor first presents a simple shape formed by a few connected vertices. A user can reshape the membership function by moving the vertices with the mouse or add new vertices to create a more complex shape.

Any variable available to CubiCalc can be plotted on a two-dimensional scrolling strip chart, a scatter chart or a polar plot. The output membership functions can also be displayed. A three-dimensional plot of the decision surface can also be presented to a viewer. CubiCalc can be used to view a surface plot with any variable as a function of two others.

Example 6.7 *New 'clicking mouse' technology of programming and software design with an ability of producing an executable software without knowing any programming language or compiling any program segments – FUZZLE™*

FUZZLE software is fully supported with graphics displays and mouse functions. One of the outputs is a source code in C or FORTRAN language that can be converted into an executable code and attached to an application environment. In addition, FUZZLE has its own execution module with graphics support which does not require any programming, compilation or linking.

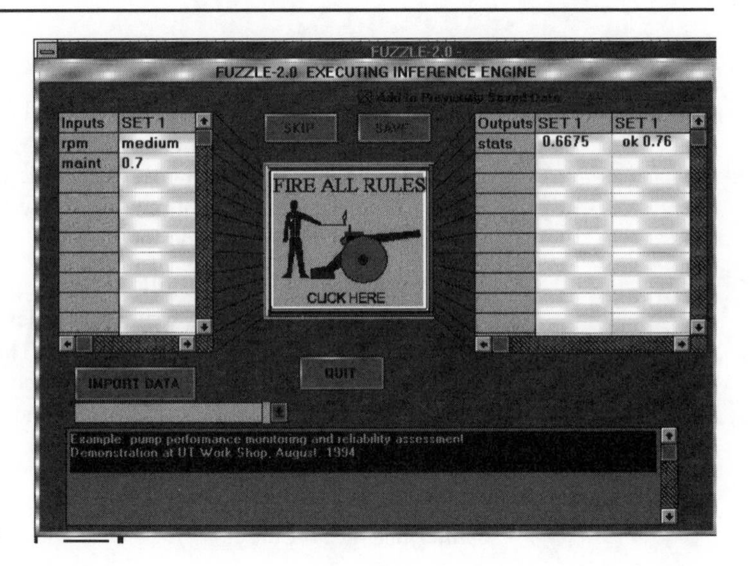

Fig. 6.4 *FUZZLE operational environment*

Using this option, the user can obtain inference results directly from the FUZZLE shell by entering data via keyboard or external data files. Once the inference engine is validated, the executable portion of the software can be extracted by a click of a button, and it can be customised for special purpose applications. The final product, which is stripped from FUZZLE related functions and screen images, is ready for distribution and it is royalty-free.

Example 6.8 *Producing an output code oriented on different hardware implementation – fuzzyTECH™*

To satisfy any user, different fuzzyTECH versions may include translators which produce an output file in C, an assembly code for INFORM's FUZZY-166 fuzzy processor, an assembly code for various Intel, SGS-Thomson, and Siemens microcontrollers, or a code for programmable logic controllers (PLC) produced by Allen-Bradley, Siemens and Klockner-Moeller. Users are supposed to purchase the different versions of fuzzyTECH depending on the type of the code they wish to generate.

Example 6.9 *Advanced simulation and debugging support – FIDE™*

FIDE provides tools for simulation and debugging separate FIUs as well as an entire system. One can debug a FIU using MF-Edit, Analyser, Tracer and Simulator.

- MF-Edit – graphically modifies membership functions. One may edit membership functions by modifying the FIL source code. However, it may be more convenient to tune membership functions graphically. The Membership Function Editor (MF-Edit) graphically defines new shapes for membership functions and saves the results in files (if possible to edit).
- Analyser – provides a three-dimensional view of the response function, it can also trace back to the source code for any anomaly point.
- Tracer – provides the user with an ability for given input values to inspect the output, inference process step by step, and trace back to the source code. Tracer is used to track specific FIU behaviour to the actual rules. Tracer also provides a detailed look at every step of the inference process. One can follow an execution from the output directly back to the rule base in the source code. The anomalies are apparent which should be addressed in the debugging process.
- Simulator – allows a user to view input and output values graphically as a function of time (steps). The simulation is driven by a set of input values which are specified by the user in a separate file. If the resulting response functions exhibit undesirable behaviour (such as the overshoot, for example), the rules and membership functions that cause this behaviour can be directly determined with Tracer, and then the FIL source code is recompiled.

These tools are used to identify and correct errors in source codes. Also, one can fine-tune one's application with other debugging tools. The Composer is a text and graphic editor used to link FEUs, FIUs and FOUs and then interconnect them. It can also be used to illustrate data flow graphically through the overall system and locate errors which may not be apparent until the entire system is analysed.

All debug utilities are able easily to identify rules and membership functions that require modification. Rule modification is accomplished using FIL editor. Membership function modification can use FIL editor or the graphical MF-Edit feature. Once changes are made, the FIU can be recompiled and the debug process can start again. The real-time code (RTC) option on the FIDE menu allows one to generate assembly code versions of one's FIU for various Motorola microprocessors.

***Example 6.10** Combining abilities with other intelligent tools – NeuroFuzzy™ Module in fuzzyTECH™*

In many applications, the desired system behaviour is represented by data sets. In control systems, these data sets may represent operational states. The NeuroFuzzy Module allows us to use these data sets for an automated fuzzy controller adjustment with any fuzzyTECH edition. An adjustment of both membership functions and fuzzy rules is supported as well as an automatic optimisation towards the data sets. The NeuroFuzzy Module (Fig. 6.5) integrates neural network technologies to train fuzzy logic systems. In contrast to conventional neural network solutions, the entire training process and the resulting fuzzy logic system remain completely self-explanatory.

The training technology used is based on a fuzzy logic inference extension that is implemented in most fuzzyTECH™ editions: the use of Fuzzy Associative Maps (FAM). Such FAM rules can be considered generalised neurones and by using the NeuroFuzzy Module technologies, neural net training methods are applicable. A fuzzy logic system generator optionally sets up appropriate membership functions and rule blocks prior to the actual training phase based on the data sets.

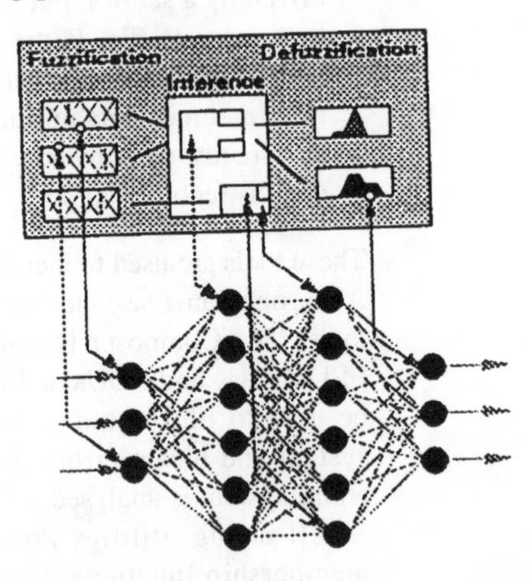

Fig. 6.5 *Neuro-Fuzzy Module structure*

The NeuroFuzzy Module can also be used to optimise existing fuzzy logic systems. Starting with an existing system, the NeuroFuzzy Module interactively tunes rule weights and membership function definitions, so that the system converges to the behaviour represented by the data sets. Any fuzzy logic system generated by the NeuroFuzzy Module can be optimised manually and verified.

Example 6.11 *Including hardware blocks together with the software tools to generate, model and simulate fuzzy controller with the embedded code – Fuzzy Microcontroller Development System by MentaLogic Systems Inc.*

The automated fuzzy controller design station (see Fig. 6.6) includes the fuzzy expert system, AutoFuzz, for generation and optimisation of automatic fuzzy rules and membership functions. It includes the FlexFuzz™ HPC1600 block for control modelling and a special evaluation board for chip development.

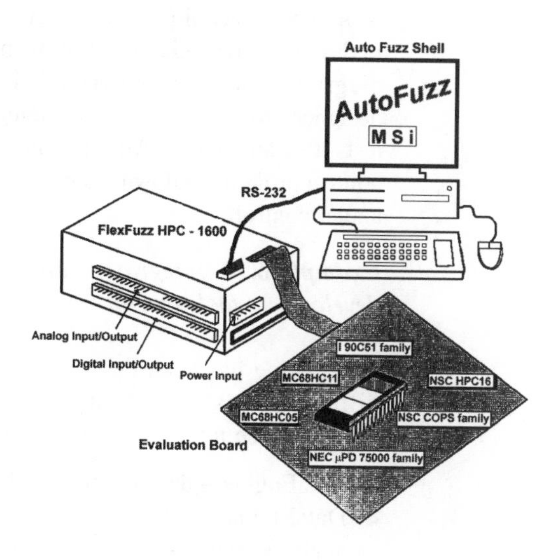

Fig. 6.6 *The structure for the design and development system*

Example 6.12 *Integration of software and hardware tools into a united complex for design, simulation and real-time control – UNAC™*

UNAC comprises two main components, GIAC™ and SAAC™. GIAC (Graphical Interface for Advanced Control) is a software package running on a host computer, usually a workstation or PC. It provides a graphical user interface for operating UNAC. SAAC (Stand-Alone Advanced controller) is a hardware unit which communicates with the host computer via both serial or Ethernet links. SAAC also communicates with the sensors and actuators of the process being controlled, via industry-standard links.

So UNAC is an integrated complex (see Fig. 6.7), combining software and hardware components, with a fuzzy control toolbox being just a part of its software half. The hardware part (SAAC)

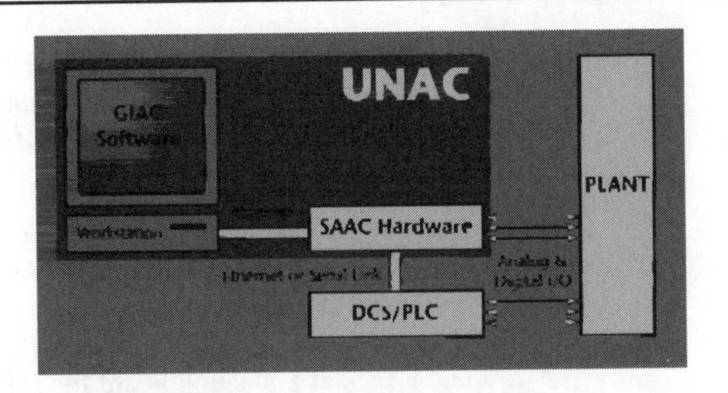

Fig. 6.7 *The integration complex structure*

is applied to acquire information from a plant. This data is used to develop a dynamic model of the process, from which the block-diagram schematic is derived. Using the dynamic model, a designer synthesises a control strategy for the process control, and tests it by simulation. After the model and control are verified, the control program is downloaded into the hardware and applied for actual control.

Example 6.13 *Integration and close interaction with other development tools and environments – DataEngine™*

DataEngine has been developed as the product which contains different tools for solving complex data analysis problems [Ang95]. The main members of this family are:

- DataEngine – development tools for data analysis;
- DataEngine ADL – C++ library for the integration of data analysis solutions in an existing software environment;
- DataEngine VI – virtual instruments for LabVIEW.

DataEngine VI incorporates the so-called Virtual Instruments to train and run neuro-fuzzy systems under the environment of LabVIEW®. So this product plays the role of an extra add-on library for LabVIEW® which enables an automation of classification and pattern recognition problems. The relationship between DataEngine VI and LabVIEW® is illustrated in Fig. 6.8 [Ang95].

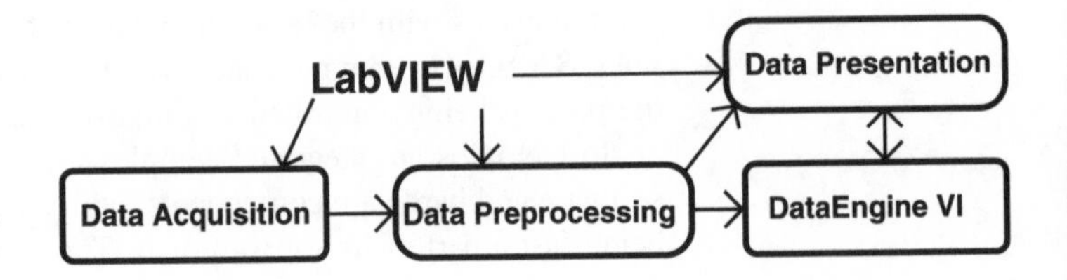

Fig. 6.8 *Integration of DataEngine VI with the LabVIEW products*

7 FUZZY CONTROLLER IMPLEMENTATION

7.1 How do we implement a fuzzy controller?

This chapter describes the hardware implementation of fuzzy controllers. It gives some advice on how to construct real fuzzy controllers. This is the most exciting part of such a developing area as a fuzzy controller design and implementation. New chips and devices are being developed and manufactured now in large quantities and any particular device will become obsolete before this book is read. So this chapter is short and filled with only general descriptions and recommendations.

The simplest and the most usual way to implement a fuzzy controller is to realise it as a computer program on a general purpose processor. However, a large number of fuzzy control applications require a real-time operation to interface high-speed external devices. For example, automobile speed control, electric motor control robot control are characterised by severe speed constraints. Software implementation of fuzzy logic on general purpose processors cannot be considered as a suitable design solution for this type of application. In such cases, design specifications can be matched by specialised fuzzy processors.

The requirements to the hardware implementation are:

- high-speed performance;
- low complexity;
- high flexibility.

These conditions contradict each other. So it is not easy to choose the right way, especially if one takes into account some other factors, such as manufacturing cost (very important for consumer product fuzzy controllers) or design cost (important in research and development).

I understand what one needs high performance for. What does low complexity mean with regard to the hardware implementation? And high flexibility?

Low complexity means that algorithms for fuzzy processing, fuzzification and defuzzification have to be very simple and demand as small an amount of memory as possible for their realisation. Flexibility means the ability of the hardware to be used successfully in different applications and configurations.

Considering these requirements, have any devices been developed yet?

During recent years, there has been an increasing interest in the development of an efficient fuzzy controller hardware capable of coping with the requirements of real-time applications. Togai and Watanabe [Tog86] developed the first fuzzy logic chip in 1985. Later Yamakawa developed a fuzzy logic hardware using analog techniques. Since then, several chips have been proposed utilising both analog and digital techniques. Generally speaking, three different ways of implementating fuzzy controller hardware can be proposed. They are summarised, together with their advantages and disadvantages, in Table 7.1.

Table 7.1

Class of hardware implementation	**Advantages**	**Disadvantages**
Digital general purpose processor	Flexibility in choice of hardware and software tools	Low performance unless a very powerful one is used
Digital specialised processor	Increasing performance	Incrementing complexity and cost, as it must be coupled with a standard host processor Lack of flexibility, as it may be applied to a limited class of problems Higher cost
Analog processor	High performance Low cost Low power consumption	Mainly the research topic Low accuracy Lack of flexibility

One can see that any hardware type has its positive and negative sides for fuzzy controller application. That's why we will briefly discuss all of them.

7.2 Implementation on a digital general purpose processor

Nowadays, most fuzzy controllers are implemented as software programs on general purpose processors and microprocessors. If there is a need for higher operation speed, a specialised fuzzy processor can be added. An example of such a processor is the FC110 from Togai.

So, if one needs a real-time operation, the specialised fuzzy processor should be applied?

Not always. The advantage of this processor is the powerful arithmetic logical unit. However, for defuzzification implementation, a fast multiplication and division are needed. If an 8- or 16-bit microprocessor is applied for a fuzzy controller realisation, the FC110 speeds up the operation significantly. If a 32- or 64-bit processor is used, the advantage of the FC110 processor is only the fast minimum and maximum operations. At the same time, other operations can be performed faster on a general purpose processor.

I understand that to speed up a fuzzy controller operation, a more powerful processor must be used.

Yes, while this is an obvious answer, it is not the only one. Other ways suppose the optimisation of the algorithm of a fuzzy controller operation.

You mean the optimisation on the target platform as proposed in [Fuz92]?

Not exactly. Some advice can be provided on how to speed up a controller operation on a general processor. In this case, the optimisation is based not on the features of a particular platform, but on the specific features of the fuzzy controller operation.

What are they?

Research has demonstrated that only two stages of a fuzzy controller operation take most of the processing time, about 83 per cent for rules processing and 16 per cent for defuzzification [Ung94]. This is why the main efforts should be and are concentrated on the inference engine implementation. On the other hand, the important features of modern general microprocessors [Sur94] should be considered as well. These features include the following:

- Addition, multiplication and division operations are about 10 to 100 times faster with integer numbers than with floating point numbers. Floating point operations are not available on microcontrollers.
- Usually minimum and maximum operations are not available.

- Jumps for short distances are fast.
- Data that is often used is held in the processor cache or register memory.

Based on this, the following recommendations can be provided:

- All calculation operations must be done with integer numbers. It can be easily realised as the input signal for the fuzzy controller comes from the AD converter which outputs an integer code.
- The membership functions for the inputs are to be stored in look-up tables. In this case the antecedent parts of the rules can be calculated very fast.
- Data dependencies, especially the property of the minimum operation, that min $(X,0) = 0$, can reduce the number of operations drastically. It can cause a significant reduction in the rules calculation time. If an antecedent part is zero and the t norm operation is interpreted as the min operation, then this rule is not activated. There is no need to calculate the consequent part. Moreover, other antecedent conditions need not be checked. All other rules with the same antecedent part do not need to be calculated either. Generally, if similar antecedent parts are used in different rules (that is quite common in fuzzy controllers), there is no need to recalculate them, and the result of the first calculation can be reused in the actual rule.

How far can the processing time be reduced with this optimisation?

Quite significantly. Let us consider a typical PD-like controller with a rules table given in Section 3.5.3. It has 49 rules with two conditions each and without any reduction $49 \times 2 = 98$ antecedent parts should be calculated. By reusing the results available, the number of antecedents to be calculated can be decreased down to 14. The number of minimum operations can be reduced as well.

Can you give us any real results?

You can find them in the literature. For example, in [Sur94] the comparison between PC80486, 33 MHz, the FC110 from Togai InfraLogic and the MC8052, 12 MHz is given. In the benchmark (see Fig. 7.1) the FC110 is nearly 20 times faster than MC8052 and needs less program memory.

In general, microcontrollers like the MC8052, which is an 8-

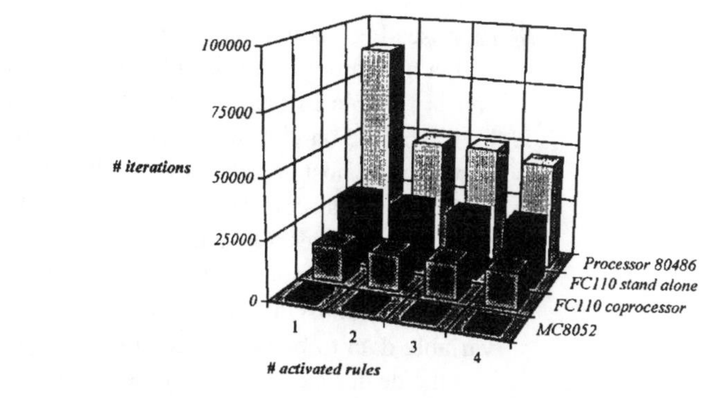

Fig. 7.1 *Comparison results for a standard microcontroller MC8052, a specialised digital processor FC110, and a general digital processor PC80486: performance benchmark in iterations per second [Sur94]*

bit type, are suitable as a cheap realisation of fuzzy controllers with a medium complexity and a low or medium execution time. These and similar microcontrollers are well tested and often used. However, the PC80486 with an optimised C-code implementation is faster than the FC110 because of the improved algorithm. By the generation and optimisation of the assembler code, the execution time for it can be reduced even further. So, for faster processing, the extension and optimisation of some basic fuzzy operations are the most important.

7.3 Implementation on a digital specialised processor

Research into the balance between performance and cost has recently led to the development of architecture solutions with a specific support, and several accelerator coprocessors dedicated to fuzzy logic and control have been proposed. Dedicated hardware may be considered as the best way in terms of performance, but it can only cover a limited range of applications. In spite of the lack of flexibility, the choice of the entire specialised hardware solutions may represent an effective way, in particular for the applications which require a large number of rules. Specific fuzzy hardware allows us in many cases to reach a better cost–performance ratio because of the exploitation of parallelism in fuzzy processing and the introduction of special purpose units.

To introduce this type of implementation, let me consider two specialised processors. The first one is FC110 from Togai InfraLogic. It is rather old already (a few years). The second one is AL220 from Adaptive Logic™, which is very new.

The FC110 digital fuzzy processor from Togai InfraLogic was developed as a specialised fuzzy coprocessor. It is a single chip small enough for sensitive embedded applications. Its architecture supposes high communication possibilities for working together with a host processor. It is not oriented to any particular host type and is flexible in possible applications. Variable data are stored in a 256 byte on-chip RAM. At least the low 64 bytes are shared between the host and the device with arbitration provided by the FC110. Special communication capabilities are assigned to two of these addresses. Off-chip data interfacing is also possible.

Due to this architecture, the microprocessor allows the program and all the constant data to reside in an off-chip ROM and the variable data to be placed in an on-chip RAM that both the host and the device can access. The shared RAM is used for temporary storage and to transfer observations, commands, conclusions. Additional RAM is provided in a 192-byte segment adjacent to the shared RAM.

The AL220 is an inexpensive, high performance, stand-alone microcontroller utilising fuzzy control. The device contains four 8-bit resolution analog inputs and four 8-bit analog outputs, and an internal clock generator. Inputs can be directly connected to sensors or switches. Outputs can be connected to analog devices or used to control a mechanism. The AL220 consumes very little power during normal operation and has a power-down mode. The AL220 diagram is given on Fig. 7.2

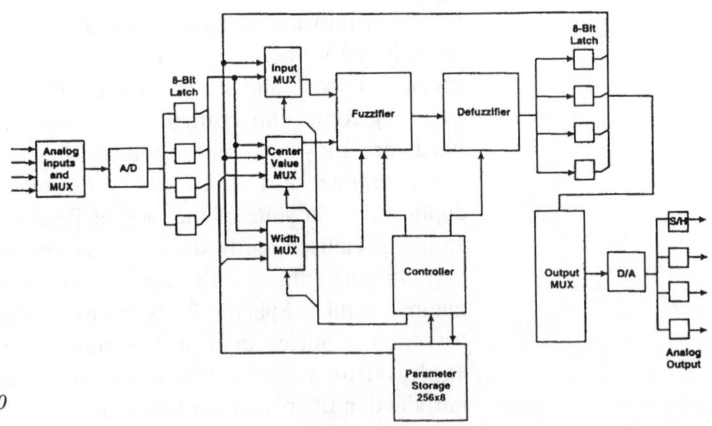

Fig. 7.2 *Detailed AL220 block diagram*

The main elements are a fuzzifier, a defuzzifier and a controller, performing fuzzy processing. The knowledge base containing rules and membership functions is realised as an EEPROM/ROM with the capacity of 256 × 8 bits. It is available in either an 18-pin DIP or 20-pin SOIC versions. One can see that this device includes on-chip A/D and D/A converters that eliminates a need in external devices and gives a designer a one-chip solution.

Can you provide more information on how it works please?

The microcontroller [Inn95] reads voltage levels from its four analog inputs using an 8-bit A/D converter, processes the channel data according to fuzzy rules contained on the chip, and generates four analog inputs via its 8-bit D/A converters and four sample-and-hold output drivers. Fuzzy processing is performed at a decision rate of 500,000 rules per second that allows one to carry out first-, second- and third-order derivatives calculation and control, automatic calibration and rule-based timing at 10,000 samples per second for each of the four analog channels.

Let us have a look at some fuzzy processor chips currently being developed. In order not to be accused of any biased approach, two fuzzy chips are presented, one of a Japanese design and another one of an European design (see Table 7.2). Data given is from the manufacturers. Omron is famous for the world's first high-speed fuzzy controller, FZ-1000. FP-3000 is a new generation fuzzy processor which is applied in different Omron products. WARP (Weight Associative Rule Processor) by SGS-Thomson (see Fig. 7.3) is claimed to be the technological state-of-the-art processor. Table 7.2 is not comparing these two chips, but presenting some real-life data and demonstrating how fast this area is being developed.

Fig. 7.3 *WARP fuzzy processor*

Table 7.2

Key features	**FP-3000, Omron**	**WARP, SGS-Thompson**
No. of rules processing inputs	8	16
No. of rules processing outputs	4	16
No. of possible membership functions for each input	7	16
No. of possible membership functions for each output	7	128
No. of types of membership functions supported	4 shapes (L,P,S,Z)	All
No. of rules	Single mode: 29 Expanded mode: 128 per group with 3 groups	Up to 256
Operation time	20 rules with 5 inputs and 2 outputs and defuzzification by centre-of-gravity method in 650 μs	32 rules with 5 inputs and 1 output are evaluated in 1.85 μs (1.5 MFLIPS)
Data resolution	Unsigned 12 bit	8 bit

You mentioned that a specialised fuzzy processor is usually applied as a coprocessor. How is this cooperation organised?

A fuzzy coprocessor by Togai InfraLogic, [Tog95], the VY86C570, is a high-performance fuzzy coprocessor with a 12-bit FCA (Fuzzy Computational Acceleration) core, 4 K × 12 OCTD (Observation, Conclusion and Temporary Data), RB (Rule Base), SMI (Shared Memory Interface), and host interface logic combined in a single chip. The VY86C570 is capable of executing simple to very complex fuzzy computations at high speeds, making it suitable for a wide range of fuzzy logic applications. Simple-to-medium complexity fuzzy logic rule bases can be directly downloaded by a host processor to 4 K words (approximately 200 rules) of on-chip rule base memory. This allows designers to create fuzzy coprocessing systems without the need for an expensive on-board memory. For larger fuzzy application requirements, the VY86C570 includes an external rule base interface that allows up to 64 K words (over 1,000 rules) of rule base memory.

In a typical fuzzy application (see Fig. 7.5) a fuzzy rule base is downloaded by the host into RB memory prior to the start of any fuzzy computation. At the beginning of a fuzzy computation, 'crisp' input values, or observations, are downloaded by the host into OCTD SMI. The fuzzy core uses the rule base information stored in the RB memory to perform calculations and produce a

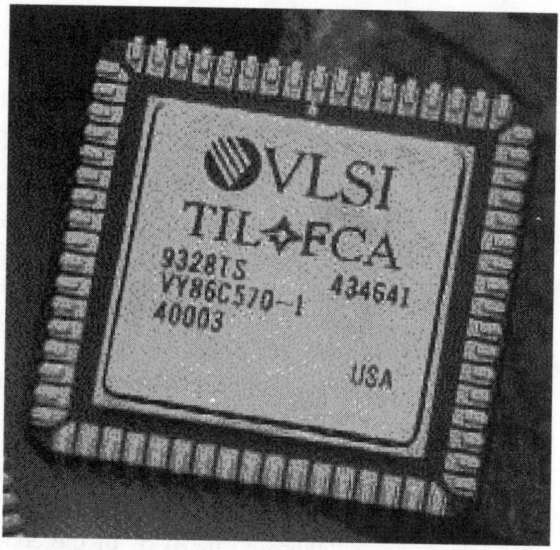

Fig. 7.4 *Specialised fuzzy coprocessor chip produced by Togai InfraLogic*

set of 'crisp' output values or conclusions. These values are stored by the fuzzy core in the OCTD SMI and are read by the host through the host interface. Control status registers are used to control the modes of the chip and to provide status read-back.

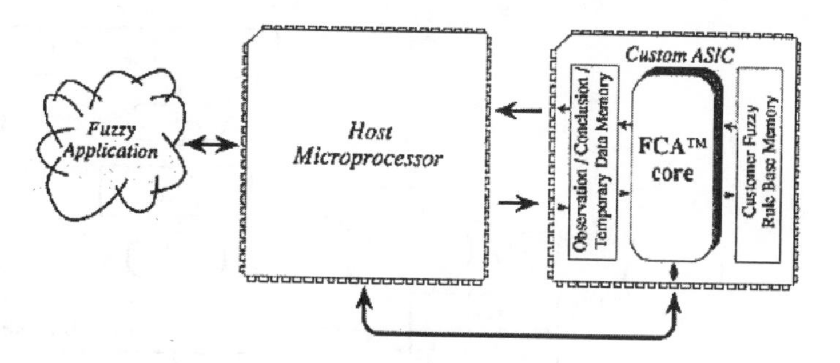

Fig. 7.5 *Typical application when a custom ASIC chip is used as a fuzzy coprocessor*

7.4 Specialised processor development system

I understand that this controller should have some rules and membership functions to fulfil all its functions. How do we fill in its electronic knowledge base?

A special design environment is supplied, to put all these data onto the microcontroller chip called a development system. FC110 has a development system, while AL220 has one called INSIGHT IIe™. This includes some software and a special hardware unit.

All together this system provides an interface and tools for design development, simulation, real-time emulation and debugging. The development process is conducted on the personal computer in an MS-Windows environment and is pretty easy to perform.

The block diagram for the FC110 development system is presented in Fig. 7.6, which explains the contents and operation of this system. In the Togai InfraLogic design package, the fuzzy controller rule base is written in FPL high level language (see Example 6.4 of Section 6.3). The development system should translate this FPL description into an executable code that can be downloaded into the FC110 processor. The system includes a special Compiler, Assembler and Linker. The Compiler translates the FPL code into the machine code optimised for the FC110 digital processor. The Assembler converts additional files written in assembler language into relocatable files. These files may contain information other than a knowledge base. The Linker combines all the parts of the code together. It utilises specific information describing the board configuration. The Linker produces the code which will be downloaded into the target board, if the knowledge base memory is implemented as RAM, burned in, if the memory is a PROM or permanently programmed into a ROM. It also outputs the C-code file which is used for the interface with the application software programs.

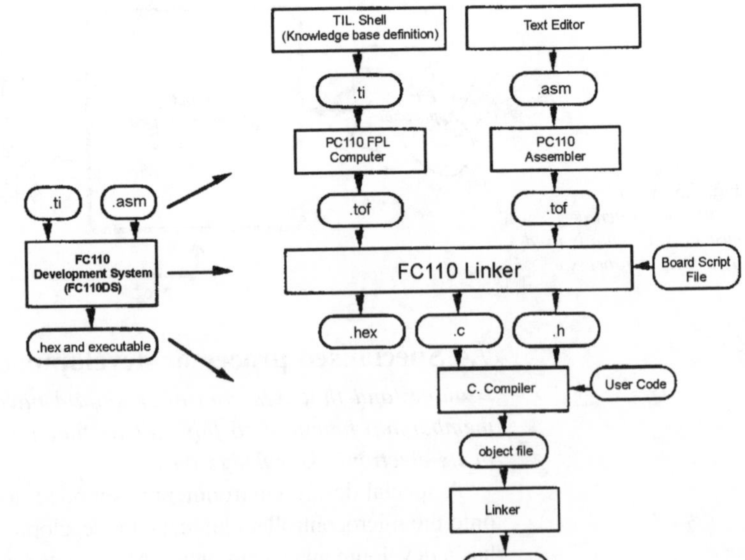

Fig. 7.6 *The FC110 development system block diagram.*

7.5 Implementation on analog devices

During the last few years, analog circuits attracted close attention as a good candidate for a fuzzy controller implementation. This implementation is characterised with a higher operation speed and a lower power consumption. The functional efficiency is also much larger than for the digital realisation because of the possibility of the versatile exploitation of small analog devices for a wide variety of low level linear and nonlinear processing required for fuzzy inference realisation.

But isn't it true that analog circuits are generally not able to provide good accuracy?

Generally, that is right. But the fuzzy controller application is a lucky exemption which does not require high accuracy. Accuracy of 6–9 bits is enough and is quite affordable even for the cheapest analog implementations. This makes analog circuits natural candidates for designing fuzzy controller chips with optimum speed-to-power ratio figures for low and medium precision applications, up to about 1 per cent. That's why some circuits and chips have been developed and implemented already.

Could you describe these chips in greater detail?

Let me give just a brief introduction [Jar94] into Yamakawa's design which is pretty representative for the hardware developed. The whole fuzzy system is divided into two parts according to their functions, that is, the rule chip for fuzzy inference (FP9000) and the defuzzifier chip for defuzzification (FP9001). This functional division facilitates flexible system configuration. The distinctive features of these chips are: high-speed fuzzy logic operation in parallel mode, compact fuzzy systems (chip saving) suitable for built-in application and adaptability of the fuzzy system based on a rule set during execution of fuzzy inference. These design features have allowed an inference speed of more than 1 mega fuzzy logical inferences per second, excluding defuzzification [Mik95].

The fuzzy engine is implemented in a parallel architecture where the consequents of all rules in response to the determined antecedents are defined and programmed internally, aggregated by an analog 'or' construction and combined to produce a defuzzified output value. The internal processing in both FP9000 and FP9001 is performed in an analog mode as opposed to other implementations, although a digital interface has been included in the latest version in order to define, modify (write) and read the parameters of each fuzzy rule quickly.

The rule chip consists of an antecedent block, a consequent block and a rule memory to store fuzzy rule sets. Up to four fuzzy rules can be stored and processed simultaneously, each with three antecedent variables and one consequent variable. The *t* norm operation applied to the antecedent parts is performed by a min circuit. To describe input fuzzy variables, only S or Z functions are allowed as membership functions. Each membership function circuit can produce up to six different alternate function types, with a total of 31 different centre positions within the universe of discourse.

The consequent block has four demultiplexer circuits (one for each rule in memory) to decode the single consequent label defined for each fuzzy rule. A three bit opcode provides seven possible combinations (code 000 is not assigned) labelled NL, NM, NS, Z, PS, PM, PL for the definition of the consequent centre values within the universe of discourse. The consequent block performs a max operation (*t* conorm). The rule memory supports a digital interface for fuzzy rules definition and application. It is a two-stage memory consisting of 24 8-bit registers (three duplicated antecedents and four inference engines) and four 3-bit registers for the consequent parts of each engine.

The defuzzifier chip accepts a set of singletons, one from each rule consequent, which are considered as the centres of the fuzzy outputs. Singletons are applied to increase the computation speed and decrease the complexity. They calculate a crisp output by the centre of gravity method, using the assigned weights for each singleton.

What are these S- and Z-type membership functions?

These are terms to describe the classes of membership given in Fig. 7.7, which are standard for hardware implementation.

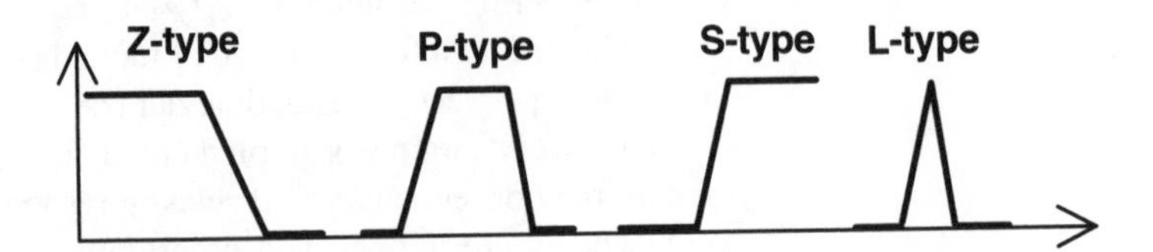

Fig. 7.7 *Standard membership function types*

How can one manage to avoid using any other types except Z-type and S-type?

One can apply a combination of these two types. For example, both P- and L-types are the minimum combination of Z- and S-type functions.

It is not necessary to apply chips to develop a fuzzy controller. One may construct the whole circuit from simple elements.

How can I do that?

Layer by layer. Because a fuzzy controller operation includes some stages like fuzzification, fuzzy rules processing, defuzzification, a circuit using it contains parts or layers corresponding to these operations. Each layer realises one of the operations. For example, let us construct the simple circuit based on the operational amplifier which realises the S-type of membership functions. This circuit is applied in fuzzification.

Note, that the Z-type membership function has two parameters: a and b. Then it can be defined as:

$$Z(a,b,x) = \min(1, \max(0, Z_0(a,b,x))),$$

where $Z_0(a,b,x) = ½ - (x-a) \times b$.

To realise this membership function the circuit given in Fig. 7.8 with the symmetric power supply V_{cc} can be proposed [Sanz94]. In this case:

$$a = R_2/R_1$$

$$b = \frac{V_{cc}(a + 1)(R_4/R_3 - 1)}{a(R_4/R_3 + 1)}$$

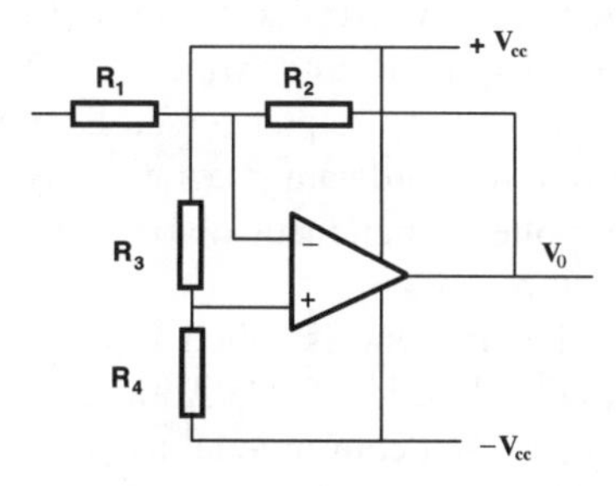

Fig. 7.8 *Circuit for the implementation of the Z-type membership functions*

Is it so simple to design an analog fuzzy controller?

It is not that difficult if one has only to consider basic design principles. The example of an analog reconfigurable fuzzy controller recently developed is given in [Guo94, Guo95]. This controller has a modular architecture and reconfigurable inference engine. It is a two-input and one-output fuzzy controller which implements Mamdani's min–max inference engine and centre-of-area defuzzification method. Each input and output includes five membership functions. The controller implements 13 rules. It was designed with analog circuits working in a voltage mode. The design was implemented in 2.4 micron CMOS technology.

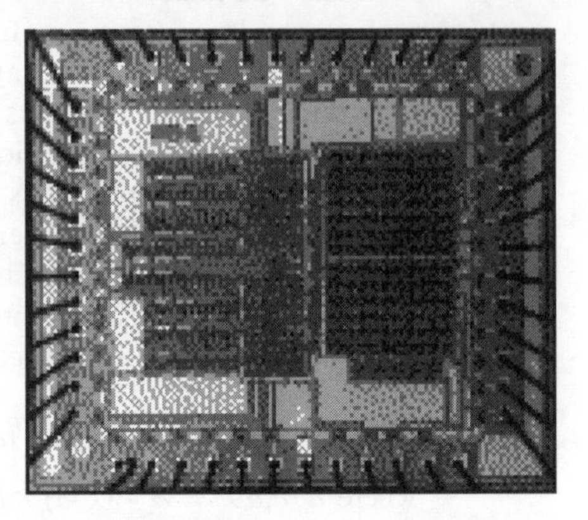

Fig. 7.9 *A reconfigurable fuzzy controller (a microphotograph)*

7.6 Integration of fuzzy and conventional control hardware

As fuzzy control has become popular and a number of successful real life applications has been developed, hardware developers have started proposing complex solutions, integrating fuzzy software and hardware with conventional PLC (programmable logic controller) and DCS (distributed control systems) hardware. The aim is to provide an opportunity to complete a design and implement a controller for any particular application without any need for any extra software and hardware. Merging of design (development and simulation) and implementation was considered in Chapter 6 (see Example 6.12). Here, attention is paid to the hardware integration. Two ways of transforming fuzzy controller design methodology into real industrial applications are given below.

The first one is called UNAC™ [CICS96, Adams95] which was developed by CICS Automation. It proposes the combination of fuzzy and conventional technology on a 'macro' level. UNAC consists of a design package, including fuzzy control methodology, and a hardware part (SAAC™ – see Fig. 7.10) which is downloaded with a controller code developed by the software. The SAAC 1000 uses a DEC Alpha AXPVME™ processor board utilising a VxWorks™ real-time operating system. It incorporates a VME backplane and supports an extensive range of VME™ cards. Communication with the workstation running a software package is via 10 Mbit/s Ethernet using TCP/IP. High-speed sampling is provided via VME based I/O cards, and other systems can be accessed using RS-232, RS-422, RS-485 and GPIB. SAAC 1000 may be integrated into DCS or PLC systems via communication protocols such as Modbus. This provides SAAC 1000 with access to the DCS/PLC operator interface and plant I/O. And in its turn, it provides the DCS/PLC system with a user-friendly advanced process control facility.

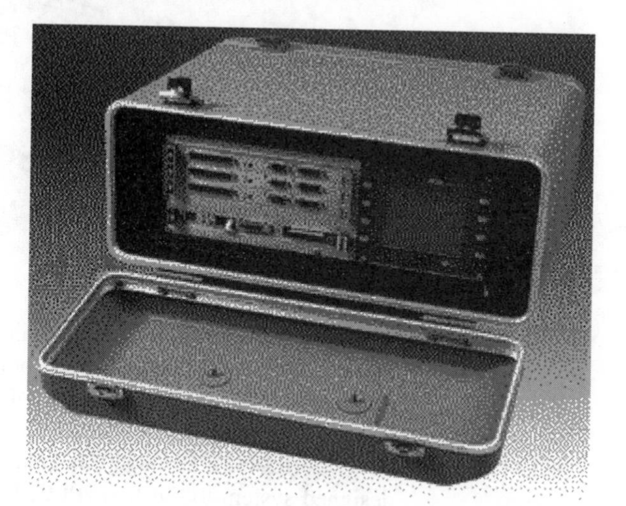

Fig. 7.10 *SAAC 1000™ portable controller*

The second methodology is the 'micro' level solution, developed by Inform Software and Klockner-Moeller, and consists of two chips, field bus connections and interfaces [Geb96]. An analog ASIC handles the analog/digital interfaces at industry standard 12-bit resolution. Snap-on modules can extend the periphery for large applications of up to about 100 signals. An integrated field bus connection, based on RS485, provides further extension by networking. The conventional and the fuzzy logic computation is handled by a 16–32 bit RISC microcontroller. The operating system and communication routines, developed by Klockner-Moeller, are based on a commercial real-time multitasking kernel. The internal RAM of 256 kB can be extended by memory cards using flash technology. Thus, the fuzzyPLC™ (Fig. 7.11) is capable of solving real complex problems of industrial automation.

The fuzzyPLC is programmed by an enhanced version of the standard fuzzy logic system development software fuzzyTECH™ of Inform Software. Unlike all other control design packages, fuzzyTECH has been enhanced with editors and functions to

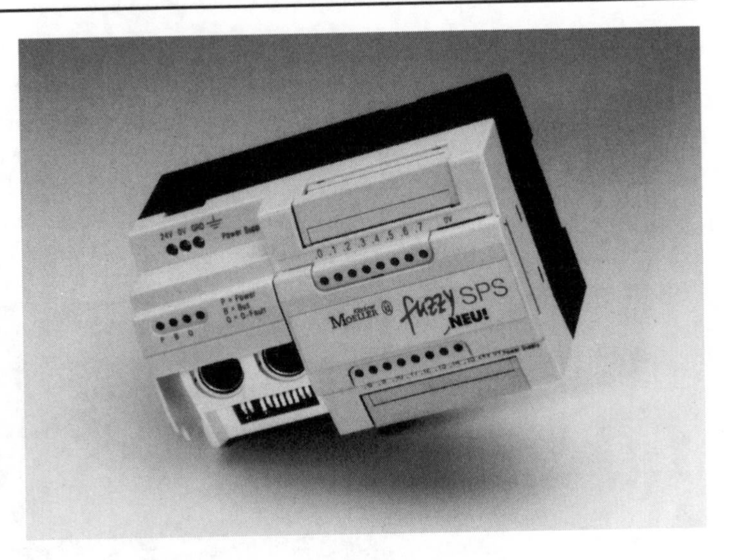

Fig. 7.11 *fuzzyPLC™, an integration of fuzzy and automation hardware*

support the conventional programming of the PLC. Thus, a user only needs one tool to program both conventional and fuzzy logic parts of the solution. The software runs on a PC and is linked to the fuzzyPLC by a standard serial cable (RS232) or the field bus (RS485). Through this link, the developer downloads the designed system to the fuzzyPLC. Because fuzzy logic systems often require optimisation 'on-the-fly', fuzzyTECH and the fuzzyPLC feature 'online debugging', where the system running on the fuzzyPLC is completely visualised by the graphical editors and analysers of fuzzyTECH. Plus, in online-debugging modes, any modification of the fuzzy logic system is instantly translated to the fuzzyPLC without halting operation.

WHAT ELSE CAN I USE? OR SUPPLEMENTARY INFORMATION FOR TEACHING AND LEARNING

8 A BRIEF MANUAL TO FUZZY CONTROLLER DESIGN

8.1 When to apply fuzzy controllers

A plant can be considered as a black box with outputs available for measurement and a possibility of changing inputs. The plant is supposed to be observable and controllable. Some information about the plant operation or plant control is available, which can or cannot be of a quantitative nature, but it can be formulated as a set of rules (maybe after some processing).

An acceptable fuzzy control solution is possible, which should satisfy design specifications. It must not be optimal in regard to some criteria as it is hard to prove that a fuzzy control system is optimal and even stable. However, a fuzzy controller is able to provide a stable and 'good' solution.

8.2 When not to apply fuzzy controllers

Fuzzy controllers should not be used when:

- conventional control theory yields a satisfying result;
- an easily solvable and adequate mathematical model already exists;
- the control problem is not solvable;

8.3 Fuzzy controller operation

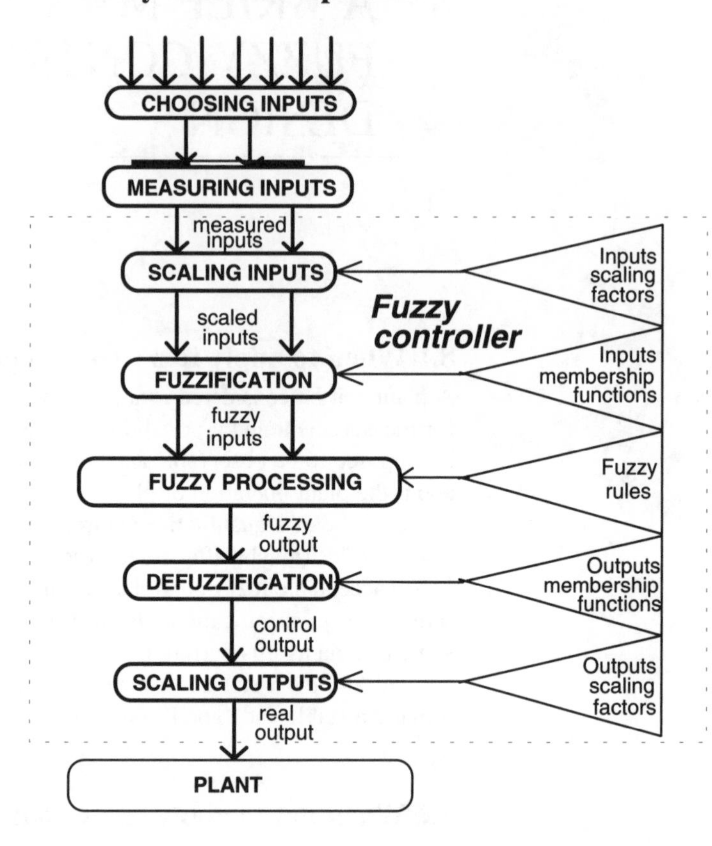

Stage 1: Fuzzification

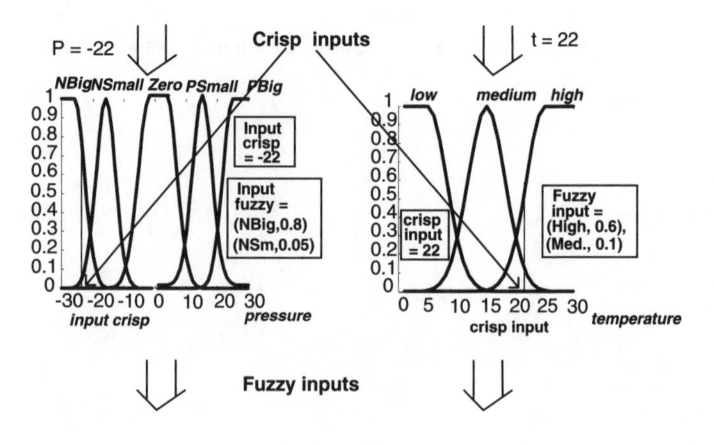

Stage 2: Fuzzy Processing

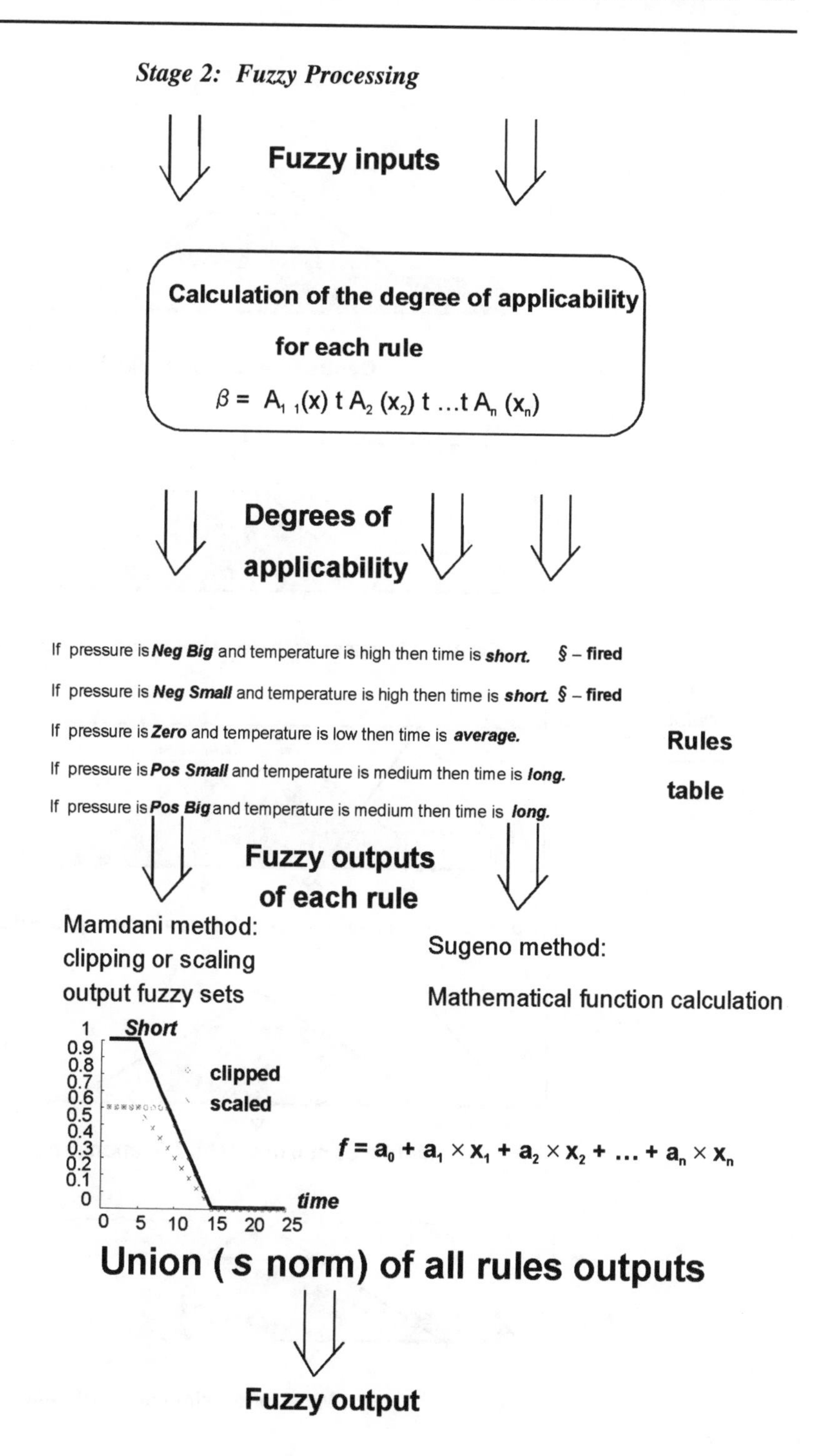

Stage 3: Defuzzification methods

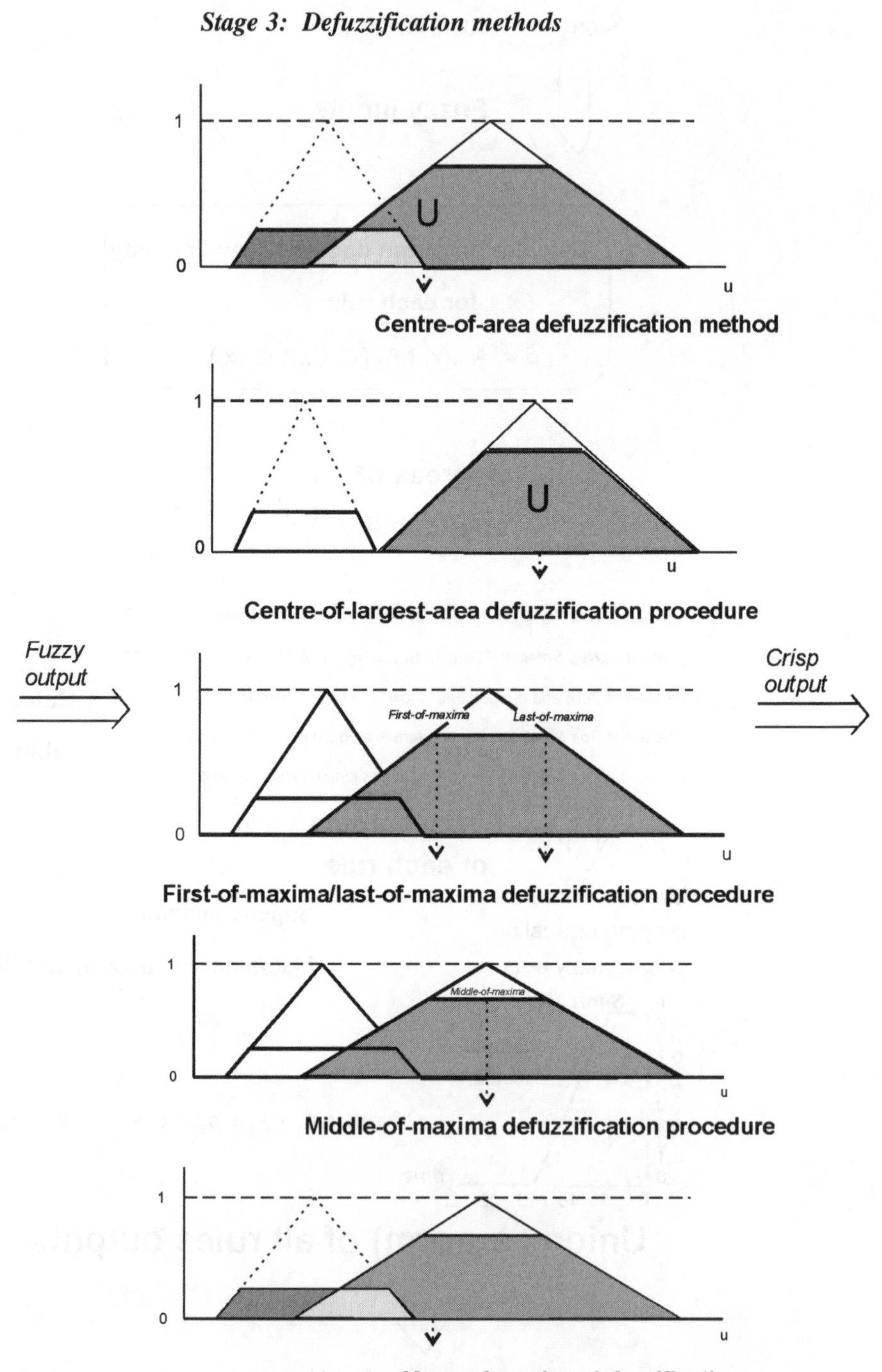

8.4 Which fuzzy controller type to choose

<table>
<tr><th colspan="3">Table 8.1 Features of different fuzzy controller types</th></tr>
<tr><th>Simple fuzzy controllers</th><th>Complex and/or multilevel fuzzy controllers</th><th>Adaptive and/or self-organising fuzzy controllers</th></tr>
<tr><td>One-level controller structure, no hierarchical rules structure</td><td>Multilevel controller structure with a few controllers at a level, hierarchical rules structure</td><td>Usually: one-level controller structure, no hierarchical rules structure
Possibly: multilevel controller structure with a few controllers at a level, hierarchical rules structure</td></tr>
<tr><td colspan="3">Fixed operational procedures including the methods of approximate reasoning, fuzzification and defuzzification</td></tr>
<tr><td colspan="2">Fixed input and output scaling factors</td><td>Input and output scaling factors can be modified (tuned) between runs or changed during the run</td></tr>
<tr><td colspan="2">Fixed inputs and outputs</td><td>Practically: fixed inputs and outputs
Theoretically: inputs or outputs can be changed</td></tr>
<tr><td colspan="2">All rules have the same degree of confidence, equal to unity</td><td>Rules may have different degrees which can be tuned</td></tr>
<tr><td colspan="2">A number of rules strongly depends on a number of inputs</td><td>Dependence is not usually so dramatic</td></tr>
<tr><td colspan="2">Fixed number of classes (membership functions) for any input and output</td><td>Number of classes can be changed (tuned)</td></tr>
<tr><td colspan="2">Fixed membership functions for each class</td><td>Parameters, shapes and position of membership functions can be changed (tuned)</td></tr>
</table>

8.5 Fuzzy controller structure and parameter choice

<table>
<tr><th colspan="2">Table 8.2</th></tr>
<tr><td>Choice of the structure</td><td>Apply the hierarchical structure whenever there is any doubt in the stability of a fuzzy control system or in applications requiring high reliability</td></tr>
<tr><td>Choice of the inputs</td><td>The same as for a conventional control system
The error and change of error (derivative) signals are often applied as the inputs for a fuzzy controller
Additional: it is easy to choose the inputs depending on some control rules, expressing the dependence of the output on these inputs</td></tr>
</table>

Table 8.2 Continued

Choice of the scaling factors	Initially choose the scaling factors to satisfy the operational ranges (the universe of discourse) for the inputs and outputs, if they are known Change the scaling factors to satisfy the performance parameters given in the specifications on the base of recommendations formulated in Section 4.3
Choice of the number of the classes (membership functions)	There are several issues to consider when determining the number of membership functions and their overlap characteristics The number of membership functions is quite often odd, generally anywhere from 3 to 9 As a rule of thumb, the greater control required (i.e., the more sensitive the output should be to the input changes), the greater the membership function density in that input region
Choice of the membership functions	The expert's approach – choose the membership functions determined by the expert(s) The control engineering approach: • initially choose the width of the membership functions to provide the whole overlap (see Section 4.4), about 12–14%; • in order to improve the steady-state error and the response time, decrease the membership function's whole overlap; • in order to improve transient characteristics (oscillation, settling time, overshoot), increase the whole overlap; • the use of a fuzzy controller with wider membership functions and a large overlap can be recommended in the presence of large disturbances.
Choice of the rules	Main methods: • expert's experience and knowledge; • operator's control actions learning; • fuzzy model of the process or object under control usage; • learning technique application. The whole rules set should be: • complete; • consistent; • continuous.
Choice of the defuzzification method	Choose the method according to the criteria (see Table 4.11) The most widely used are: the Centre of Area and Middle of Maxima
Choice of the fuzzy reasoning method	Choose Mamdani method if the rules are expected to be formulated by a human expert Choose Sugeno method if computational efficiency and convenience in analysis are important

Table 8.2 Continued

Choice of the t norm and s norm calculation method	See Section 2.2 The most widely used are: • for t norm min or product operators; • for s norm max or algebraic sum.

8.6 How to find membership functions

Table 8.3

	Method	Source of information applied	Brief procedure description
Subjective approach	Expert's professional knowledge and intuition	Based on an expert's ability to generate information through his/her knowledge and understanding of the problem and/or problem area	An expert assigns the membership degree to any element of the universe or proposes the whole function for the fuzzy set definition. A fuzzy set can be described with the linguistic term
	Opinion poll results processing	The same as previous but applied to a set of experts and followed by processing to get the results	As previous but the degrees assigned by individual experts are processed altogether to get total characteristics
	Ranking	Based on expert(s) ability to compare and rank different objects	Expert(s) assess prefer ences, often by pairwise comparisons, determining the membership degrees by this way
	Logic inference	Based on deductions from available knowledge (nature laws, expert knowledge, etc.)	Membership degrees are deduced from some information available and related to the object considered
	Inductive reasoning	Based on deriving membership degrees from particular facts (data sets)	Membership degrees are derived by generalising some available data
Objective approach	Fuzzy statistics	Based on statistical processing of the data available	Membership degrees are derived by the methods of mathematical statistics
	Control engineering	Based on assignment of membership functions from recommendations of control theory	Membership functions are assigned according to some rules derived from control theory methods

Table 8.3 Continued

	Method	Source of information applied	Brief procedure description
Objective Approach	Neural networks	Based on modelling membership functions or their parameters with neural networks	Neural networks become a part of a neuro-fuzzy system modelling, for example, membership functions
	Genetic algorithms	Based on choosing parameters of the membership functions with genetic/evolutionary algorithms	Parameters of the membership functions initially chosen are changed by applying a special optimisation technique

8.7 How to find rules

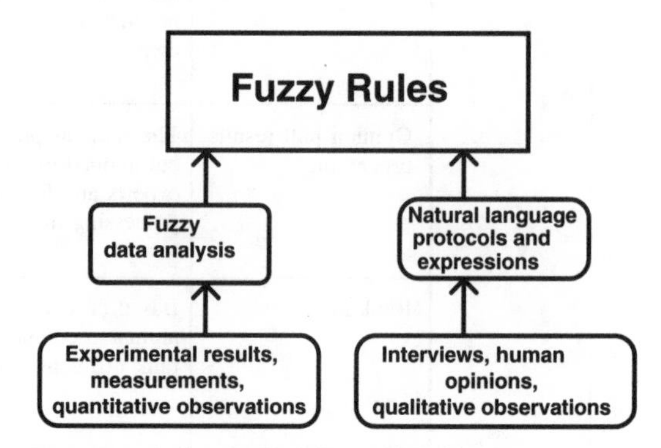

Fig. 8.1 *Main sources and ways of rules formulation*

8.8 How to implement a fuzzy controller

Table 8.4 Fuzzy controller implementation

Conditions	Recommended implementation
Simple fuzzy controller with a rather small number of rules for applications with a high production rate (e.g. consumer products)	Analog circuit processor
Complex fuzzy controller with a large number of rules for real-time industrial applications (e.g. robot control, motor control)	Digital specialised processor
Fuzzy controller for research applications, test development, and other applications without special requirements	Digital general purpose processor

8.9 How to test a fuzzy controller

Because the theory of fuzzy controller analysis has just started to be developed, the most widely used method of testing is by simulation. The first and most important criterion which the fuzzy controller must satisfy is common sense and one's professional knowledge and the knowledge of an expert in the application domain.

Some advice is as follows:

- It is better to test a fuzzy controller jointly with the plant, if the plant or its model is available.
- Use one of the design packages available to simulate the fuzzy controller and the plant.
- During simulation, try different combinations of the inputs and observe the corresponding outputs.
- Try values for the inputs which you suspect may 'break' the fuzzy controller, such as:
  - input values at the extreme ends of the universe of discourse;
  - input values at the extreme ends of the individual membership function domains;
  - input values corresponding to membership function overlap.

Many design packages allow for a control surface creation. Analyse the control surface obtained. You must verify that the control surface conforms to your expectations. A lack of conformity is usually due to one of three problems in this order:

1. Incorrectly specified rules.
2. Incorrectly defined membership functions.
3. Inappropriate fuzzy operators.

Make sure that the strong changes of the surface (maximum rate of change of input vs output) are not concentrated on the borders for all surface views. If they are, you may want to:

- re-examine and possibly modify the scaling factors;
- Make the membership function density higher in the area of maximum surface changes – this will tend to make the slope more graceful.

Check to see if the surface has unexpected anomalies. A common reason for this is that rules refer to the wrong labels.

8.10 How to fix a fuzzy controller

Table 8.5 Fuzzy controller design troubleshooting

What's wrong	What to do
Fuzzy controller does not provide stability to the system	1. Change the structure of the control system. Apply the supervisory fuzzy controller. However, you need to remember that this approach usually decreases the controller performance to some degree. 2. Check the rules table: sometimes one needs to change an incorrect sign in the rules output. 3. Decrease the output scaling factor of the fuzzy controller.
Overshoot (output oscillation magnitude) is too high	1. Decrease the output scaling factor of the fuzzy controller. 2. In a PID-like fuzzy controller, decrease the output scaling factor of the PD-part.
Speed of response is too low (rise time is too long)	1. In a PID-like fuzzy controller increase the output scaling factor of the PD-part. 2. Increase the scaling factor for a differential input compared to other inputs.
Poor steady-state accuracy	1. In a PID-like fuzzy controller, decrease the output scaling factor of the PD-part. 2. Increase the scaling factor for an integral input compared to other inputs. 3. Apply the centre of gravity defuzzification method. 4. Reduce the width of the membership function for the zero class of the error signal. 5. Redistribute the membership functions, increasing their concentration around the zero point.
Insufficient sensitivity to the input signal	Increase the scaling factor for this input.
Anomalies in the control surface	1. Check the rules table: it may contain some incorrect rules. 2. Check the rules table against required properties. 3. Check the membership functions distribution: they must have the right overlap.
Insufficient speed (performance)	1. Try another defuzzification and fuzzy inference method. 2. Consider another implementation (a specialised digital or analog processor).

8.11 How to choose a design package

Table 8.6

Company	Products	Goal	Platform	Environment	Features
Adaptive Logic (USA) Stopped operations product is available from other companies: *see* Nu Horizons, 10 Maxess Rd, Melville, NY11747. Fax: (1516) 396 5050	INSIGHT AL220	An easy-to-use interface for design, development, simulation, real-time emulation, and debugging of fuzzy microcontroller AL220 Fuzzy microcontroller	IBM PC compatible	MS-Windows	Design entry is performed using select-and-click, pull-down menus, results are viewed graphically See Chapter 7
American NeuraLogix, Inc (USA) 411 Central Park Dr. Sanford, FL 32771 USA Fax: (1407) 322 5609	NLX110 NLX230	Fuzzy pattern comparator Single chip fuzzy microcontroller 8- and 16-bit VLSI core elements for fuzzy processing			Chips described as fuzzy devices but probably are not actually realising fuzzy technology
Aptronix Inc. (USA) 2040 Kington Pl., Santa Clara, CA 95051, USA. Fax: (1408) 4902729 http://www.aptronix.com/	FIDE	Fuzzy inference development environment	IBM PC compatible	MS Windows	Allows to edit, simulate, debug and tune the membership functions and rules for fuzzy system applications. The debug tools allow time domain simulation, 3D surface displays
Byte Craft (Canada) 421 King Str N. Waterloo, Ontario N2J4E4 Canada Fax: (1519) 746 6751 http://www.bytecraft.com/	Fuzz-C C6805	A stand-alone C preprocessor for fuzzy logic	IBM PC compatible	MS-DOS or PC-DOS	Accepts fuzzy logic rules, membership functions and consequence functions and produces C-code

Table 8.6 continued

Company	Products	Goal	Platform	Environment	Features
Bell Helicopter Textron (USA) TSIE, Inc. P.O. Box 14155 Albuquerque NM 87191 USA http://www.bellhelicopter.textron.com/home.html	FULDEK	Fuzzy logic development kit – a CAD package written in visual BASIC	IBM PC compatible	MS-DOS or PC-DOS	Allows the user to execute the resulting fuzzy inferences. Many options of defuzzification and encoding are provided so that the user can find the best fit for an application
CICS Automation (Australia) P.O. Box 570 Wallsend, NSW 2287, Australia Fax: (612) 7965 6705 http://www.cicsanto.com.au	UNAC	Prototyping and implementing advanced process control algorithms, diagnostic studies and signal analysis	Includes a software package and a hardware controller		Fuzzy control toolbox provides a graphic user interface, simulation and testing of the fuzzy control systems. When completed, the design is implemented directly on the process
Flexible Intelligence Group, LLC (USA) P.O. Box 1477 Tuscaloosa AL 35486 USA http://www.flextool.com/ftfs.html	FlexTool A modular software tool which provides an environment for applying fuzzy systems to diverse domains.	A Version M2.2 is a MATLAB version which provides an environment for design in MATLAB	Various	MATLAB environment	Combines the uncertainty handling capabilities, qualitative reasoning capabilities of fuzzy systems with the evolutionary learning capabilities of genetic algorithms
FRIL Systems Ltd (UK) Bristol Business Centre, Maggs House, 78 Queens Rd. Bristol BS8 1QX UK	FRIL	Logic-programming language for solving uncertainty problems	Various		Includes an opportunity to manipulate fuzzy and probabilistic uncertainties, enabling different forms of uncertainty to be integrated within a single framework

Table 8.6 continued

Company	Products	Goal	Platform	Environ-ment	Features
Fuzzy Systems Engineering (USA) 12223 Wilsey Way, Poway CA 92064 USA Fax: (1619) 748 7384	Fuzzy Knowl-edge Builder, Fuzzy Decision Maker, and Fuzzy Thought Amplifier	Support and assistance in design fuzzy systems, modelling real systems, generating and tuning fuzzy rules	IBM PC compat-ible	MS-Windows	Helps in making the decisions, provides an interface for capturing expert's judgements
FUZZYSOFT AG and GTS Trautzel GmbH (Germany) Gottlieb-Daimler-Str. 9 D-24568 Kaltenkirchen, Germany Fax: (49) 4191 88665	FS-FUZZY-SOFT	A complete fuzzy logic based operating system which assists in developing complex software	IBM PC compat-ible	MS-Windows and MS C/C++ or Borland C/C++	Includes: graphic simulation without the need for compilation, system integration, structural and modular editing, efficient C-code generation
FuzzyWare (USA) P.O. Box 11287 Knoxville Tn 37939 USA Fax: (1-615) 588 9487	FuziCalc	A spreadsheet which deals with fuzzy uncertain numbers while retaining a conventional spreadsheet use	IBM PC compat-ible	MS-Windows	Allows to put in any cell a precise number, a fuzzy number, or a range of values. Includes financial and mathematical functions
HIWARE (Switzer-land) Tel: (41-61) 331 7151	HI-FLAG	Fuzzy development environment	IBM PC compat-ible	MS-Windows	Supports most of the MOTOROLA's microcontrollers
HyperLogic Corp. (USA) P.O. Box 300010 Escondido, CA 92030 USA Fax:(1619) 746 4089	CubiCalc, CubiCalc RTC, CubiCalc RuleMaker, CubiCard, and CubiQuick	Generating, writing, and verifying fuzzy rules	IBM PC compat-ible	MS-Windows	Provides cost-effective programming support for popular C and C++ compilers

Table 8.6 continued

Company	Products	Goal	Platform	Environ-ment	Features
Grymer Worm Tongue Inc. (UK) 199 Askew Rd. London UK http://www.compu-link.co.uk/~the-zone/assist.html	Assist Visual Designer	Fuzzy knowledge based system generator	IBM PC compat-ible	MS-Win-dows	
ICCT Technologies (Canada) Suite 205,500 Alden Rd. Markham, Ontario L3R 5H5 Canada Fax: (905) 475 5350 http://www.stria.ca/icct	FuzzyC++ FuzzyShell	A software package including a fuzzy language, compiler A software design and development pakage	IBM PC compat-ible	MS-Win-dows	GUI design, code generation for C/C++ and Hex, self-organisation of fuzzy rules and sets, adaptive fuzzy computational algorithms, support for fuzzy FPGA/ ASIC chips
Indigo Software (UK) 25 Clarence Sq. Cheltenham Gloucester-shire GL5O 4JP UK Fax: (44-01242) 243225 http://www.indigo.co.uk	FuzzyExpert– a C++ source code library for embedded fuzzy applica-tions	Includes all code one needs to delay fuzzy reasoning in one's application	Various plat-forms	C++	Features the library of C++ codes to create applications with fuzzy logic. Includes ultra-fast fuzzy expert system
Inform (Germany) Pascalstrasse 23 D-5100 Aachen Germany Fax:(49 2408) 6090 http://www.fuzzytech.com	fuzzyTECH	Development and design tools for fuzzy system design	Various		Provides all tools for fuzzy system design: online technology for modifications of the running system, all-graphical and other editors, multiple graphical analyser and

Table 8.6 continued

Company	Products	Goal	Platform	Environment	Features
					tracers, NeuroFuzzy and DataAnalyser modules, etc.
Integrated Systems (USA) 3260 Jay Str. Santa Clara, CA 95054 http://www.isi.com	RT/Fuzzy in MATRIXx	Practical tool for fuzzy systems modelling and simulation in an integrated environment	UNIX work-stations		Easy-to-develop online applications, possibility of producing embedded applications with the C-code generation, rapid prototyping of the control systems
KentRidge Instruments (Singapore) Blk 51 Ayer Rajah Ind Est Ayer Rajah Cres. Singapore 0513 Fax:(65) 7744695	FlexControl	Rapid prototyping software for fuzzy control systems	IBM PC compatible	MS-Windows	Three powerful graphical tools: fuzzy rule editor, fuzzy membership processor and fuzzy graphical analyser; fuzzy rules and membership functions can be modified interactively in real-time implementation to tune a system response

Table 8.6 continued

Company	Products	Goal	Platform	Environment	Features
Logic Programming Associates Ltd. (UK) St.4, RVPB, Trinity Rd, London SW18 3SX, UK Fax: (44 181) 874 0449 http://www.lpa.co.uk	FLINT	Fuzzy logic inferencing system that makes fuzzy technology (fuzzy variables, modifiers, rules, defuzzification) within a sophisticated programming environment (PROLOG)	IBM compatible PC or UNIX stations	MS-Windows, Mac or DOS + PROLOG	Applied to program fuzzy expert systems Allows the inclusion of fuzzy rules, modifiers, variables and defuzzifiers into an expert system
Management Intelligenter Technologien (MIT) GmbH (Germany) Promenade 9 52076 Aachen Germany Fax: (49 2408) 945 82	DataEngine PENSUM	Fuzzy and neuro data analysis and process control Modelling complex control systems using fuzzy Petri nets	IBM compatible PC or SUN-Sparc stations UNIX-station	MS-Windows	Family includes: • DataEngine – development tool for data analysis • DataEngine ADL – library for the integration of data analysis solutions into existing software • DataEngine V.i. – virtual instrument for LabVIEW, has the advantage of more human-like control
The Math Works (USA) 24 Prime Park Way Natick, MA 01760 USA Fax: (1508) 653-6284 http://www.mathworks.com/iee	Fuzzy Logic Tool Box for MatLab	A powerful tool for fuzzy system design, analysis, and simulation in an integrated environment	Various	Various	Allows creation and editing of fuzzy inference systems either by hand, with interactive graphical tools or command-line functions, or by generating automatically with clustering or adaptive neuro-fuzzy techniques

Table 8.6 continued

Company	Products	Goal	Platform	Environment	Features
MentaLogic Systems (Canada) 145 Renfrew Dr. Markham, Ontario L3R 9R6 Canada Fax: (1-905) 940 0321	Quick-Fuzz, Auto-Fuzz, Multi-Fuzz, Flex-Fuzz, Process-Fuzz, Auto Sim-Fuzz, Fuzz-Drive, F-MTOS, CT-FLC	A large number of different products aiming at: fuzzy controller design, applications development, fuzzy knowledge generating, simulating, and implementing		MS- or NT-Windows	Specialises in the development of advanced fuzzy logic technologies
Metus Systems Group (USA) 1 Griggs Lane Chappaqua NY 10512 USA	Metus library	A library of fuzzy processing modules for C/ C++			
MODiCO, Inc. (USA) P.O. Box 8485 Knoxville TN 37996-0002 USA Fax;(1-423) 584 4934 http://www.modico.com	FUZZLE – fuzzy systems development shell which generates programs in different languages and has its own execution module	Software allowing the programming of an inference engine, execution, output in C or FORTRAN code, editing and analysis. Able to create rules from data	IBM PC compatible	Various languages	Module that executes inference engines in a graphical environment without compiling or linking source code Mixed input processing capability that allows numerical input data entry as well as linguistic Analyse performance of the rules

Table 8.6 continued

Company	Products	Goal	Platform	Environment	Features
Motorola Inc. Apply to the nearest dealer	Fuzzy Logic educational kit	The computer-based tutorial set includes: fuzzy logic and control fundamentals, fuzzy system design and implementation	IBM PC compatible	MS-Windows	A new version includes FUDGE which generates assembler or ANSI C code for Motorola's microcontrollers
National Semiconductor (USA) 2900 Semiconductor Dr. PO Box 58090 Santa Clara, CA 95052 USA http://www.nsc.com	NeuFuz4	Neural-fuzzy technology for control applications which uses backpropagation to select fuzzy rules and memberships			Includes learning kits and development systems and software packages for design and developing neuro-fuzzy systems, fuzzy rules and membership functions
Omron Corp. (Japan) 4-7-35 Kitashinagawa, Shinagawa-ku, Tokyo 140 Japan Fax: 03-5488-3273 http://202.32.89.iy	Fuzzy Inference Software FS-30AT, FP-1000, FP-3000, FP-5000, E5AF	Comprehensive fuzzy inference tools	IBM PC compatible, FP-3000, FB-30AT	MS-DOS	Definable rules and membership functions, simulations, training exercises Fuzzy microcontrollers and chips, temperature process controller
Parallel Performance Group (USA) Fax: (1-520) 774-0896 http:/www.ppgsoft.com/ppgsoft/oifin.html	O'INKA	Design framework for building intelligent systems	Various platforms		Allows for integration of fuzzy logic, neural networks and user-defined modules in a single framework. Combines graphic user interface, design validation, simulation and debugging, C-code generation and design documentation

Table 8.6 continued

Company	Products	Goal	Platform	Environ-ment	Features
SGS-Thomson Microelec-tronics http://www.st.com contact a dealer	WARP processor and software develop-ment tools	Development software tools for WARP fuzzy processor	IBM PC compat-ible	MS-Win-dows	Includes editor, compiler, debugger, simulator and exporter for designing fuzzy controller, generating code and testing applications
Siemens AG (Germany) Attn: W.Keim Siemens AG ANL 441-VE G.Scharowsky Str. 2 91052 Erlangen Germany Fax: (49 9131) 732127 E-mail: siefuzzy@zfe.siemens.de	SieFuzzy	Software development tool for designing, simulation, testing fuzzy systems, compiling any resulting C-code	Various plat-forms		Interfacing with standard software packages (Matlab and Mathematica), automatic optimisation of fuzzy systems, user exten-sions (for version 2.0)
Syndesis Ltd (Greece) Iofondos 7 Athens GR-116 34, Greece Fax: (30-1) 7257829 E-mail: syndesis@-hol.gr	FLDE – an application generator which allows a designer to develop and test embedded fuzzy applica-tions	Fuzzy logic design tool producing an ANSI C-code for embedded applications	IBM PC compat-ible	MS-Win-dows	Ability for a user to achieve an optimal code generation in terms of memory and precision by choosing from a wide range of C data types. The C-code generated is self-contained

Table 8.6 continued

Company	Products	Goal	Platform	Environment	Features
TecQuipment Ltd (UK) Bonsall Str. Long Eaton, Nottingham NG10 2AN England Fax: 44(0) 1159731520 E-mail: sales@tec quip.co.uk	CE124 Fuzzy Logic System – a combined hardware and software tool for learning about and implementing fuzzy systems	Self-contained teaching aid for students to explain fuzzy logic using analogue fuzzy system elements complemented by fuzzy software	IBM PC compatible		Includes software package enabling fuzzy control to be implemented on an IBM compatible PC. The open design allows a wide range of fuzzy logic and control experiments to be developed and extended to suit the user's needs
Togai InfraLogic (USA) 18 Technology Dr. Suite 146 Irvine CA 92718 USA Fax: (1714) 588 3808	TILShell+ and other products	Software development tool which provides a way to describe fuzzy systems, test them through simulation and compile an implementation code	Various	Various	Includes fuzzy programming language, object editors and compilers
Wolfram Research Inc. (USA) 100 Trade Center Dr. Champaign IL 61820-723 USA Tel: 1-217-398-0700 Fax: 1-217-398-0747	Fuzzy Logic Pack	Creating, modifying, and visualising fuzzy sets and systems	Available for all platforms that run Mathematica 2.2	Mathematica 2.2 or later	

9 PROBLEMS AND ASSIGNMENT TOPICS

Chapter 1

Choose any one of the following topics and write a short (10–15 pages) essay.

1. What I like in fuzzy logic is ...
2. What I do not like in fuzzy logic is ...
3. Fuzzy sets theory was introduced by Lotfi Zadeh in 1965. Why not earlier or later?
4. Probability vs fuzziness: Which is the winner?
5. In 1975, William Kahan pointed out some contentions to fuzzy logic:
   - I cannot think of any problem that could not be solved better by ordinary logic.
   - What Zadeh (and fuzzy logic) is saying is the same sort of thing as: 'Technology got us into this mess and now it cannot get us out.'
   - What we need is more logical thinking not less. The danger of fuzzy theory is that it will encourage the sort of imprecise thinking that has brought us so much trouble.

Choose any of these statements and try to prove or refute it.

6. What is wrong in fuzzy sets theory?
7. Present your understanding of the concepts of complexity, uncertainty, ambiguity, imprecision, vagueness. Where is the place for fuzziness among these concepts?
8. Fuzzy theory is often determined as an instrument to model uncertainties. Do you agree with this statement? Why?
9. Historically, a long time before fuzzy theory was put forward by L. Zadeh, a probability theory had been considered as a primary tool for representing uncertainties, which is why some researchers look at fuzzy theory as a rival to probability. What is your position in this argument?
10. Propose a new device or a method where fuzzy technology could be implemented.

11. Propose a plan and ideas for a fuzzy systems laboratory.
12. A fuzzy logic educational kit: ideas, proposals, realisation.
13. Fuzzy control and classical control: what is the advantage of fuzzy control?
14. What I do not understand in fuzzy logic is ...
15. What do you consider as the most important part of fuzzy theory and technology?
16. Describe a possible application of fuzzy technology in car design.
17. Propose some ideas for an automatic car driver implementation based on fuzzy technology.
18. Propose a brief plan of an intelligent superhighway implementation based on fuzzy technology.

Chapter 2

1. Give some examples of crisp sets.
2. In the examples given in problem 1, is it easy to determine for any element of the universal set whether it belongs to a particular set or not?
3. For the sets A and B, pictured in Fig. 2.3, determine

   $$\overline{A}, \overline{B}, \overline{A} \cap \overline{B}, \overline{A} \cup \overline{B}$$

4. Determine the complement:
   - to the set of all odd numbers;
   - to the set of all numbers which are multipliers by 5.
5. Determine the intersections of:
   - the multipliers by 4 and the multipliers by 3;
   - the multipliers by 5 and the multipliers by 6;
   - the multipliers by 2 and the multipliers by 8;
   - the multipliers by 4 and the multipliers by 6.
6. Right or wrong?
   - the set of horses which won the races is a subset of all racing horses;
   - the set of white cats is a subset of all cats older than two years;
   - the set of all values for room temperature is a subset of all possible values for room temperature.
7. What is the universal set for the subset of pens?
8. What is more general, a crisp set or a fuzzy set? Can you consider a crisp set as a special case of a fuzzy set or, vice

versa, a fuzzy set as a special case of a crisp one?

9. Right or wrong?

   - if the universal set has a finite number of elements, its crisp subset has a finite number of the elements;
   - if the universal set has a finite number of elements, its fuzzy subset has a finite number of the elements.

10. Right or wrong?

    - every fuzzy set can be considered as a subset of the universal set;
    - the universal set is another fuzzy set;
    - a fuzzy set is completely determined by its membership function;
    - a fuzzy set is the union of all its level sets;
    - a membership function is the union of all its membership degrees;
    - a fuzzy set is a crisp set of pairs (u, $\mu(u)$), where $u \in U$ is an element of the universe and $\mu(u)$ is its membership degree.

11. Determine the intersection and the union of the sets of:

    - all men taller than1.8 m and all men shorter than 1.9 m;
    - all women older than 25 and all men younger than 18;
    - of red pencils and blue pencils;
    - of green pencils and all pencils longer than10 cm.

12. Draw the membership functions for the fuzzy sets of:

    - high indoors temperatures;
    - good indoors temperatures.

13. Based on the results of Problem 12, construct the membership functions for:

    - not high indoors temperatures;
    - good or high indoors temperatures;
    - 0.5-level set of the fuzzy set of high indoors temperatures.

14. Right or wrong?

    - the height of a fuzzy set is determined by the element on its crossover level;
    - 0.2-level set has more elements than 0.5-level set.

15. Draw the membership functions for the fuzzy sets of:

- low atmospheric pressure;
- high blood pressure.

16. Based on the results of Problem 15, construct the membership functions for:

    - not high blood pressure,
    - low atmospheric pressure and high blood pressure.

17. Propose membership functions to determine fuzzy sets of ultrasonic sounds and infrasonic sounds. What universe can be used to determine these fuzzy sets?
18. Depending on age, a person can be called a baby, a child, a teenager, an adult. Determine membership functions to describe these fuzzy sets.
19. After a traffic accident, the witnesses have described the speed of one vehicle involved as 'very high' and of another one as 'rather high'. Propose the membership functions to determine these fuzzy sets.
20. In the graphic presentation of the membership functions describing fuzzy sets of 'well educated people' and 'poor educated people', which unit would you propose to scale the universe?
21. α-cut can be considered as an example of a way to convert a fuzzy set into a crisp one. What is this conversion necessary for?
22. Right or wrong?

    - operations on fuzzy sets can be determined in any way one wants;
    - operations on fuzzy sets have one strict definition;
    - both the union and the intersection operations can be determined through the triangular norm;
    - the triangular norm is a two-place function.

23. Based on Definition 2.8, propose a new function to realise t norm.
24. Propose another problem similar to the one considered in Section 2.2 (see Fig. 2.11) to illustrate a possibility of using different ways to calculate the t norm.
25. Consider two fuzzy sets:

    long pencils = {pencil1/0.1, pencil2/0.2, pencil3/0.4, pencil4/0.6, pencil5/0.8, pencil6/1},

    medium pencils = {pencil1/1, pencil2/0.6, pencil3/0.4, pencil4/0.3, pencil5/0.1}.

Use two methods to calculate t norms. Determine the results of the union and the intersection of these fuzzy sets. Compare these results.

26. Apply two fuzzy sets with the membership functions given in Fig. 2.4b. Use two methods to calculate t norms. Determine the results of the union and the intersection of these fuzzy sets. Compare these results. What can you say about the dependence of the result membership function shapes on the t norm calculation method used?

27. Which ones of the following functions could be recommended as the membership functions for the fuzzy sets 'about 5', 'much higher than 10', 'low':

$$\text{a)} e^{-\left(\frac{x-5}{3}\right)^2}; \text{b)} \int e^{-x^2}; \text{c)} (10-x)^2; \text{d)} 1-0.1(x-5)^2; \text{e)} arctn(x-10); \text{f)} 1/tn(x-5)$$

28 Let the fuzzy sets A, B and C be the subsets of the universe {1,2,3, . . . , 199, 200}.

Let A = {1/0.1,2/0.3,3/0.3,4/0.4,5/0.5,6/0.7,7/0.8,8/0.9, 9/1,10/1} and
B = {1/0.1,2/0.3,3/0.5,4/0.7,5/0.9,6/1,7/0.8,8/0.5,9/0.2, 10/0}. Find out the fuzzy sets

C = A + B;
C = A * B;
C = (A + B)*A;
C = A*(A + B).

29. Under the conditions of problem 28, find the fuzzy sets

$C = A^2$;
C = 1/B.

30. Calculate the results of the following operations with the intervals:

- [1,2] + [2,4]
- [–3,4] – [5,6]
- [–3,4] × [–2,3]
- [7,8] / [2,4]
- [7,8] / [–4,–2]

31. Which of the functions below could be considered reasonable to serve as a membership function for the fuzzy set 'medium' in control applications?

$$\text{a)} \frac{x}{x+5}; \quad \text{b)} 5^{-(x+4)}; \quad \text{c)} \frac{1}{1+10(x-5)^2}$$

32. Actual resistance values of any real resistors are known to be different from nominal ones. Can this actual value be considered as a fuzzy set?
33. Propose membership functions to describe 'a fuzzy resistor' with a nominal value of 1 MΩ and another one of 5.6 kΩ.
34. Determine a fuzzy equivalent resistance for the circuit given below where R1 and R2 are fuzzy resistors described in Problem 33.

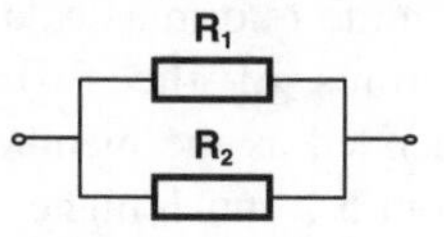

Chapter 3

1. What is the difference between Mamdani-type fuzzy controller and Sugeno-type fuzzy controller?
2. Which method of reasoning is the most appropriate model of human thinking?

   - Boolean logic;
   - fuzzy logic;
   - semantic networks.

3. Right or wrong?

   - fuzzy controller operation is based on rules processing;
   - fuzzy rules processing results in a crisp output, which is applied to an object or process under control;
   - a rules table uses measured physical variables as its inputs;
   - fuzzy processing consists of choosing one of the rules and calculating the output corresponding to this rule.

4. Describe fuzzy processing and calculate its output for Example 3.1 considered in Section 3.2.3 with the pressure error of –7 kPa and temperature 12°.
5. Suppose the fuzzy rule has two conditions in its antecedent part. Two real inputs match the membership functions corresponding to these conditions with the degrees of 0.3 and 0.7. Calculate the applicability degree of these inputs to this rule.
6. Let the fuzzy controller have three rules. Each rule has two conditions in the antecedent parts and two inputs. Two real inputs match the membership functions corresponding to the conditions of the first rule with the degrees of 0.3 and 0.7, of the second one, 0 and 0.9, and of the third one, 0.4 and 0.5. The antecedent parts of the rules are:

- $y = 0.4x_1 + 0.8\ x_2$;
- $y = 0.1x_1 + 0.9\ x_2$;
- $y = 0.2x_1 + 0.5\ x_2$;

where y is the controller output. Calculate the controller output corresponding to these inputs.

7. What stages can fuzzy controller operation be divided into?
8. What does a knowledge base of a fuzzy controller contain?
9. What scales can be proposed for vacuum cleaner fuzzy controller inputs? What scale can be used to describe a surface type? What scale can be used for the output if a possible sucking force range is up to 100 N?
10. What are the typical inputs and outputs:
    - for a PD-like fuzzy controller?
    - for a PI-like fuzzy controller?
    - for a PID-like fuzzy controller?
11. Table 9.1 contains rules for a PD-like fuzzy controller. Fill in empty places.

Table 9.1

Δe \ e	NB	NM	NS	Z	PS	PM	PB
NB		NB				NS	Z
NM			NM				
NS	NS			Z			
Z					Z		
PS			Z				PS
PM		Z			PM		
PB	Z		PS				PB

12. Suppose Table 9.2 contains rules for a PI-like fuzzy controller. Fill in empty places.

Table 9.2

Δe \ e	NB	NM	NS	Z	PS	PM	PB
NB	PB		PB			PS	Z
NM			PM				
NS		PS		Z			NS
Z					Z		
PS			Z				
PM					NM		
PB	Z			NS			NB

13. Right or wrong?
    The structural difference between PD-like fuzzy controllers and PI-like fuzzy controllers is concentrated mainly in:

    - choice of inputs;
    - choice of outputs;
    - choice of rules.

14. If all the rules are to be formulated by experts, what type of fuzzy processing can be recommended?
15. What functions are usually applied in the consequent part of the rules in Sugeno fuzzy processing?
16. What type of fuzzy processing should be chosen for the fuzzy controller designed for a robot mobile application?
17. Which defuzzification method can be recommended for the Sugeno-type fuzzy controller with two inputs and 25 rules designed for a process control system?
18. Give the definition of controller stability. How does a stable controller differ from an unstable one?
19. What is the advantage of a hierarchical fuzzy controller over a simple one?
20. Determine which ones of the rules given below represent a fuzzy rules base.

    - **If** temperature is below 15° **then** switch on a heater for 5 minutes.
    - **If** temperature is low **then** switch on a heater for 5 minutes.
    - **If** temperature is below 15° **then** switch on a heater for a few minutes
    - **If** temperature is low **then** switch on a heater for a few minutes

    where Low means below 16° and a few means between 3 and 5.

21. In a multilevel hierarchical structure, which level works under more uncertain conditions, high or low?
22. Do you agree with the statement that a fuzzy controller is an applied fuzzy expert system?
23. Formulate some rules for the fuzzy controller controlling the movement of a car approaching a stop sign. What inputs should this controller have?
24. What type does the controller designed in Problem 23 have?

Chapter 4

Fuzzy control has the potential to change the way people and goods are moved around a factory, a neighbourhood, city, country. Steering a moving vehicle, controlling the speed at which a vehicle travels, avoiding obstacles and target tracking are key aspects where fuzzy logic could be implemented to reduce the work on humans. By removing the need for humans to worry about breaking the speed limit, they can concentrate more on where they are going and what is in their way, and hopefully that will reduce the number of deaths and accidents on the roads each year. This set of problems tends to help in developing fuzzy controllers for an intelligent super highway.

1. We know that when a set of traffic lights is green, the car can travel through the intersection, if the light is yellow, then the driver must either stop or hurry through the intersection. Write down the corresponding rules set for a fuzzy controller.
2. Take into account the time, how long the current light has been fired. How should the previous set be modified?
3. Now try to design a fuzzy controller for traffic lights. Which inputs should be used for this controller?
4. Now design a fuzzy controller to stop a car at the traffic lights. Which inputs should you use? How should you formulate the rules?
5. Parking is some drivers' worst nightmare. Write down the rules for a fuzzy controller performing reverse parking.
6. On the future intelligent superhighway, all cars will be separated into different palettes, which are groups of cars travelling with the same speed. The distance between palettes as well as the distance between cars within a palette should be kept constant during travelling. Determine the main difference between two fuzzy controllers: the first one controlling the distance between the palettes and the second one controlling the distance between cars within a palette.
7. What sensors will be required to realise the fuzzy controllers mentioned above?
8. Propose a rules table for a PI-like fuzzy controller for a room temperature control.
9. Take a manual for any control device. Design a rules table based on this manual.
10. The fuzzy control system developed for an automatic flight control of the X-29 aircraft consists of three PID-like fuzzy controllers: for pitch, roll and yaw control [Luo95]. What is the main difference between these controllers?

11. In each fuzzy controller described in Problem 10, the rules set consists of coarse and fine rules. The coarse rules are designed to supply fast response with a large control input, while the fine rules are designed to make a fine adjustment to improve a dynamic stability. Is it a hierarchical structure? If so, what is a lower level?
12. Table 9.3 contains the protocol results for some computer communication sessions. The fuzzy system to be designed has to evaluate the communication quality. It may consist of rules like:

    If errors are seldom and repeats are seldom **then** quality is high.

    Which inputs and outputs would you like to apply to this fuzzy system?

Table 9.3

Session	A	B	C	D	E	F
Packets sent	551	52	37	187	50	1185
Packets received	1861	1234	97	34	129	4717
CRC errors	0	0	2	1	0	1
Retransmits	153	48	232	194	232	141
Timeouts	50	98	68	435	68	42
Failed sanity checks	1	0	0	0	0	0
Incomplete packets	1	1	1	2	1	1
Bytes received	608 305	894 567	640 195	427 456	648 979	1584 727
Bytes discarded	0	0	0	0	0	0

13. Can a number of retransmits in Problem 12 be applied as an input?
14. Add some other rules to the fuzzy system of Problem 12.
15. A control engineer often evaluates a controller quality by analysing its response. What inputs for this fuzzy evaluation system could you propose?

16. Develop some rules for the fuzzy evaluation system of Problem 15.
17. Which membership functions, Fig. 9.1a, b, c or d would you prefer to apply to model a fuzzy controller error input and why?

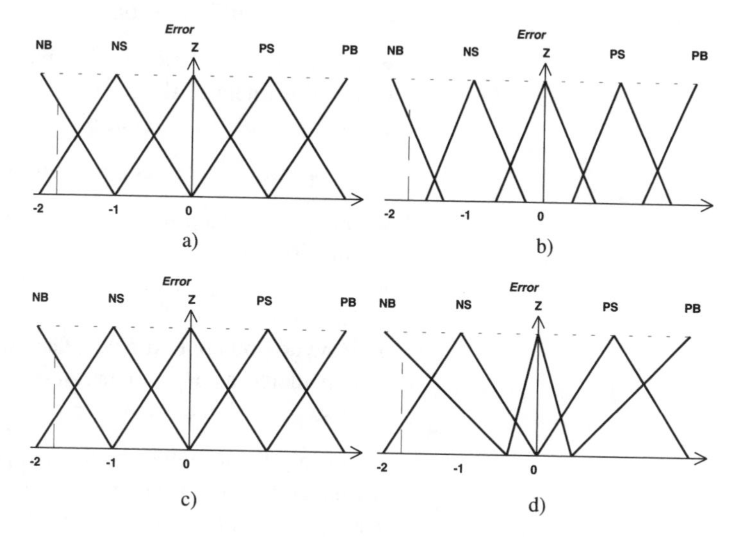

Fig. 9.1. *Membership functions*

18. The fuzzy controller has two inputs: error (e) and change of error (Δe), and its membership functions are presented in Fig. 4.25. This controller is given by the rules set of Table 9.4. Is this rules set complete? Why?

Table 9.4

e \ Δe	NB	NS	Z	PS	PB
NB	PB		PB		Z
NS	PB	PB		Z	
Z			Z		NB
PS	PS		NS	NB	
PB	Z	NS		NB	

Chapter 5

1. What are conventional and non-conventional methods for fuzzy controller parameter adjustment?
2. What is the main reason for an ANN application in a fuzzy controller design?

3. What are the similarities and differences between neural network and genetic algorithms techniques?
4. What is the difference between genetic algorithms and evolutionary algorithms?
5. Which one of the methods, genetic or evolutionary algorithms, could be recommended for the following:

   - an initial formulation of the rules;
   - an adjustment of the membership function widths;
   - tuning of the fuzzy controller output scaling factors.

6. ANN operation depends on nonlinear activation functions.

   - Show that a step function is a special case of a sigmoid function.
   - Is a ramp function a special case of the sigmoid function?
   - How do you approximate a Gaussian function? Can it be considered as a combination of two sigmoid functions? If so, in which case?

7. According to [Kung93] the most popular neural model is a two-layer back propagation network. Give two reasons why.
8. The response of a system controlled by a fuzzy controller demonstrates a large overshoot. Which of the fuzzy controller scaling factors should be adjusted and how?
9. If the steady state error value is unacceptable, how would you adjust the parameters of the fuzzy controller?
10. Will the adjustment recommended in Problem 9 change other fuzzy controller characteristics?
11. In the process of choosing the scaling factors for the X29 aircraft flight fuzzy controller [Luo95], the scaling factors are first estimated by evaluating the largest possible value of each variable and are then adjusted to give a better performance in terms of quick response and small tracking errors. What do you think of this procedure? What does the stage of an initial choice include?
12. In an ANFIS system [Jang95], which part of the fuzzy controller is realised by the ANN?
13. [Wu96] starts with the statement that 'fuzzy control is a direct method for controlling a system without the need of a mathematical model, in contrast to the classical control which is an indirect method with a mathematical model.' Do you agree with this definition?

Chapter 6

1. How would you propose to improve a fuzzy controller design package which you applied?
2. How many 'bugs' have you discovered working with the design package?
3. What are the main features of the fuzzy controller design package?
4. What was the main reason for including the neuro-fuzzy module into a fuzzy design package?
5. Why did large companies wish to produce their own fuzzy design packages? What did these packages include?
6. How is a typical fuzzy design package organised?
7. What level of a human–machine interface is usually achieved in a fuzzy design package? How can you prove your conclusion?
8. What type of programming is required from a user of a design package?

Chapter 7

1. In a hardware implementation,
   - when are analog circuits preferred over digital circuits?
   - when are digital circuits preferred over analog circuits?
   - when are hybrid circuits preferred over either analog or digital circuits?
2. Can you propose a way to improve the performance of a fuzzy processor?
3. Which stages of a fuzzy controller operation must be optimised first of all? Why?
4. Can a fuzzy controller work in a real-time regime? What does it depend on?
5. What is a fuzzy coprocessor applied for?
6. How is the joint operation of a general purpose digital processor and a fuzzy coprocessor organised? Which operations should be performed by each processor?
7. Briefly describe the design of any fuzzy coprocessor chip available on the market. What are the performance characteristics of this chip?
8. What is a fuzzy processor development system for? How is it applied?
9. Which types of membership functions are usually applied in fuzzy system hardware implementation? How do you explain this choice?

10. In Fig. 7.8, an electrical circuit diagram for Z-type membership function realisation is given. What can be modified in this circuit to make it realise the S-type?
11. Propose an electrical circuit to realise a P-type membership function.
12. What precision can be achieved in an analog implementation of fuzzy controllers?

10 DESIGN PROJECTS

This chapter discusses some possible topics for student projects. Consider the following topics as examples only. The topics presented here could be useful for specialists and students with various backgrounds: engineering (electrical, computer, mechanical, aerospace), information technology, computer and mathematical sciences.

Projects 1 and 2

The problem description is given in Section 4.1. The plant simulation model is given on Fig. 4.1. Change initial conditions in blocks 7, 16 to zero. At the starting point, these initial conditions set the vessel heading directed straight to the destination point.

1. Design a fuzzy controller to move the vessel along the straight line with as small a deviation as possible. Simulate its operation and the vessel movement with a design package. Observing a vessel trajectory, evaluate the controller quality.
Try two versions of the controller:

- using a deviation from the straight line as an input;
- using the deviation angle from the target direction as the input.

Try two inputs PI-like and PD-like fuzzy controllers. Compare the trajectories with the previously observed ones.

Suggest a possible further improvement of the controller quality.

2. Design a fuzzy controller to make a vessel perform a U-turn and then move along a straight line. The required time for completing a turn should be as small as possible. To simulate this vessel movement, change the initial condition in block 16 of the plant model (See Fig. 4.1) to 180.

Project 3. Temperature-compensated VCO

In Section 3.6.4 (Example 3.7) a fuzzy controller design for power system synchronisation and stabilisation was considered. A similar problem is typical for signal generators of all types, which must compensate for the frequency drifts caused by temperature changes. Usually discrete solutions combined with voltage controlled oscillators (VCOs) are used. VCOs provide variable frequency control by varying the excitation voltage level. However, a discrete circuitry with a temperature sensor input is required to provide a DC voltage offset to compensate for frequency drift. This circuitry, composed of a variety of potentiometers and variable capacitors, requires calibration, a process in which just one system can often take hours to calibrate.

The AL220 PAIC™ (Programmable Analog IC, manufactured by Adaptive Logic) can replace discrete compensation circuitry used with VCOs. The circuit diagram designed is given in Fig. 10.1. In the temperature compensation application, the AL220 operates as a look-up table. Voltage regions are defined for the temperature sensor input and are used to control the voltage offset level output. The voltage regions are variable and can change in size from region to region and from system to system. Each voltage region maps to a specific voltage offset value for each output through the rule set. The offset values are based on the temperature characteristics of the circuitry being compensated.

The example algorithm developed by Adaptive Logic is in Fig. 10.2.

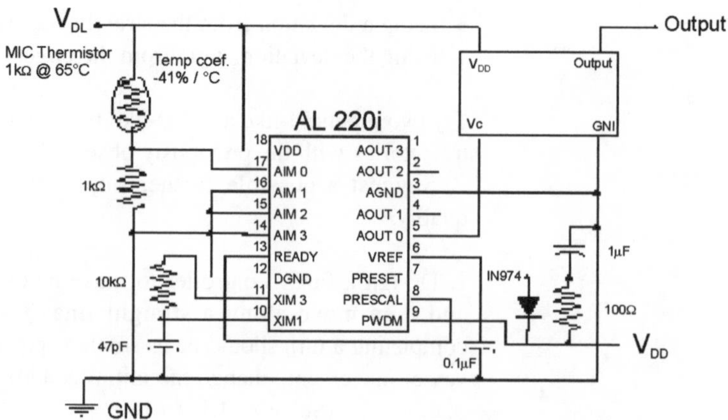

Fig. 10.1 *Temperature-compensated VCO. Example circuit schematic developed by Adaptive Logic, Inc.*

The controller has just one input and one output.

Inputs: TEMP
Outputs: VCO

Inputs regions:
TEMP is –30...–25
TEMP is –25...–20
TEMP is –20...–15
TEMP is –15...5
TEMP is 5...20
TEMP is 20...30
TEMP is 30...35
TEMP is 35...45
TEMP is 45...60
TEMP is 60...70
TEMP is < 62
TEMP is < 63.5
TEMP is 64...66
TEMP is > 66.5

The following rules have been proposed.
Rules:

If TEMP is –30...–25 **then** VCO = 137.
If TEMP is –25...–20 **then** VCO = 133.
If TEMP is –20...–15 **then** VCO = 130.
If TEMP is –15...5 **then** VCO = 128.
If TEMP is 5...20 **then** VCO = 127.
If TEMP is 20...30 **then** VCO = 129.
If TEMP is 30...35 **then** VCO = 132.
If TEMP is 35...45 **then** VCO = 130.
If TEMP is 45...60 **then** VCO = 136.
If TEMP is 60...70 and TEMP is < 62 **then** VCO = 141.
If TEMP is 60...70 and TEMP is < 63.5 **then** VCO = 147.
If TEMP is 60...70 and TEMP is > 68.5 **then** VCO = 174.
If TEMP is 60...70 and TEMP is > 66.5 **then** VCO = 161.
If TEMP is 60...70 and TEMP is 64...66 **then** VCO = 153.

Fig. 10.2 *An algorithm for temperature drift compensation*

The proposed temperature compensation algorithm is an open loop control. It operates as a look-up table. One rule adjusts the output voltage for each voltage region defined from the temperature sensor input. Voltage region selection is based on the temperature characteristics of the VCO and the frequency deviation limits.

Determine if the proposed controller is a fuzzy controller? What should one do to redesign it as a fuzzy controller? Make necessary changes to get a fuzzy controller. Could the same circuit be used to realise this controller?

Project 4. Sewer level control

In 1992, a sewer in the main street of the Melbourne suburb of Kensington (Australia) collapsed leaving a large hole in the middle of the road. For a few weeks, untreated sewage went into the bay causing much public concern. The old sewer was made of bricks. Over time, as the sewer level went up and down, the mortar between bricks was loosened, which caused the collapse. To prevent further damage to the sewer, it was important to keep the level in the sewer constant.

Four 45 kW pumps and piping were installed to bypass the collapsed part of the sewer. Two of the pumps were driven by a variable speed drive (VSD) while the other two were fixed speed. Initially operators adjusted the frequency of the VSD to try to keep the sewer level constant. However, due to the fluctuating flow of sewage into the sewer, this was unsatisfactory. The human operators did not have a quick enough response.

The Hartmann & Braun PID controller was applied then. A 4–20 mA level signal was looped into the controller. A level set-point was entered into the controller by an operator. The controller acted to increase the frequency of the VSD if the level went above the set-point and to decrease the frequency of the VSD if the level fell below it. During commissioning, the self-tuning function was tried, but it did not work properly. At first just proportional control was applied. The proportional gain was adjusted until the response was more or less satisfactory. Integral control was then added to decrease the steady-state error. Derivative control was tried but this made the response worse.

Design a fuzzy controller to control the VSD. Determine the structure of the controller, its inputs and the rules table. What type of a fuzzy controller should be chosen?

Project 5. A pumping station control

Fresh water must be pumped from a reservoir to an elevated tank several kilometres away. Three pumps are to be installed and controlled by a fuzzy controller. The control strategy includes the rule that if the water level in the elevated tank drops below 7.9 m, a pump would start. If the level drops below 7.5 m, a second pump would start and if the level dropped below 7.0 m, a third pump would start. Once the level reaches 8.7 m, a pump would stop. At 9.0 m, a second pump would stop, and at 9.2 m, all pumps would stop.

As the water pressure varies over a large range, it is necessary for each pump to have a variable speed drive to achieve the desired flow rate. The water pressure varies due to changes in:

- reservoir water level;
- elevated tank water level;
- consumer demand for water in the region.

The last influence can be modelled as a random variable depending on time of the day and season.

Design a fuzzy controller and simulate its work. How many fuzzy controllers should be applied in this system? Can a fuzzy controller be applied for solving this control problem despite crisp limits for the levels?

Project 6. A tank level control

A controller has to keep a constant level in the second tank (see Fig. 10.3). The system is nonlinear with a dead time, so it is not easy to design a conventional controller. Let us denote the levels in the tank 1 and tank 2 as h_1 and h_2, respectively, the cross-sectional areas as D_1 and D_2 and the cross-sectional areas of each pipe as A_1 and A_2. L corresponds to the dead time. Then the plant model can be described with the following equations:

$$\dot{h}_1(t) = \frac{q_1(t-L) - q_2(t)}{D_1},$$

$$q_2(t) = A_1\sqrt{2gh_1(t)},$$

$$\dot{h}_2(t) = \frac{q_2(t) - q_3(t)}{D_2},$$

$$q_3(t) = A_2\sqrt{2gh_2(t)}$$

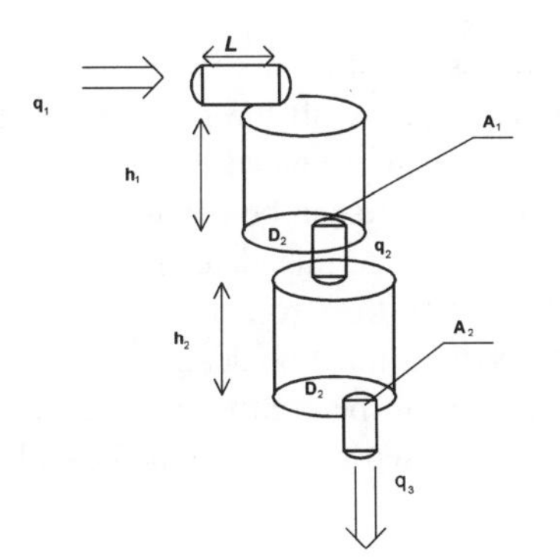

Fig. 10.3 *The two-level tank plant model*

The controller goal is to keep the level h_2 at a constant value of the setpoint. In [Tan95] a fuzzy phase-lead compensator is proposed to control this plant.

Design a fuzzy PI-like controller. Choose some parameters initially and then adjust them with some adjustment technique. Try a fuzzy PD-like controller. Compare the results.

Project 7. Control of gas and air supply to a temperature control system

The problem of keeping a constant temperature in a single room (thermostatic chamber) has been investigated for a very long time. The standard solution includes an application of a heater with a regulated gas burner (Fig. 10.4) for a water boiler. The amount of heat is controlled by a regulation of gas and air supply. The thermostatic chamber is equipped with a temperature sensor. The control goal is not only to keep the given temperature but also to provide a highly efficient operation of the heater. That means the control strategy has to aim at energy saving and environment protection by reducing the amount of gas and air supplied to the minimum quantities required for a perfect combustion in the burner. So the problem is the multicriterial control problem.

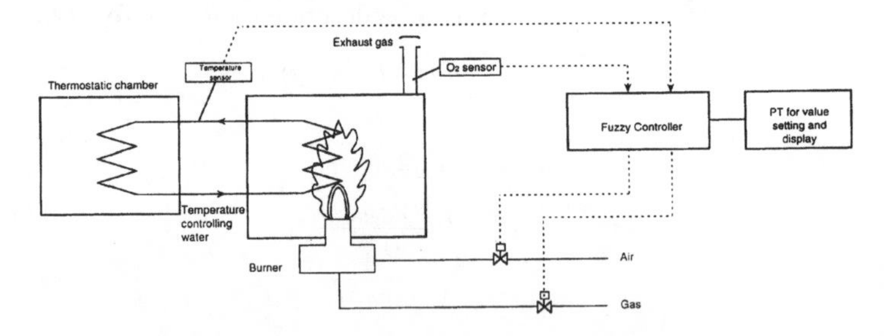

Fig. 10.4 *Fuzzy control for a heater burner*

The proposed solution was developed by Omron Corp. The burner is equipped with the oxygen sensor which measures the rate of the oxygen produced in the exhaust gas. Based on expert knowledge, this rate can be calculated for any given situation. In the project version A, this ideal percentage is supposed to be given by the expert. In the project version B, the fuzzy rules (fuzzy inference system) must be designed to derive it. Both gas and air valve drives are to be controlled with separate fuzzy controllers which are proposed to be designed as PD-like fuzzy controllers (Fig. 10.5).

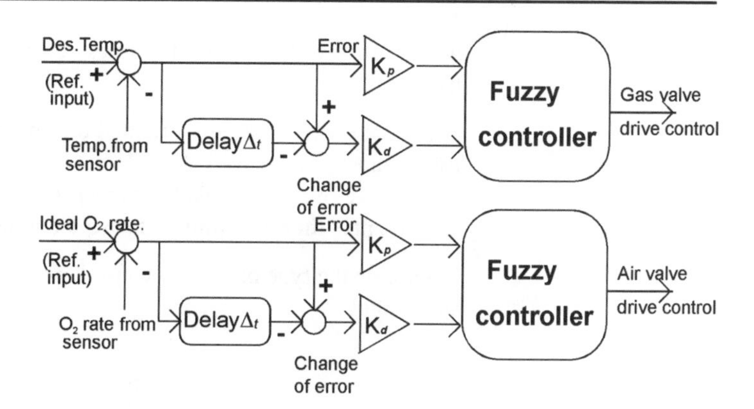

Fig. 10.5 *Proposed structure for a burner control system*

Design two PD-like fuzzy controllers controlling valve drives. The inputs for the controllers should be the differences between the desired and actual temperature in the chamber and the ideal and actual percentage of oxygen in the exhaust gas.

Project 8. Camcorder with image stabiliser problem description

In a video camcorder (see Fig. 10.6) one of the main control problem is an image stabilisation against trembling. Here the problem of separating large changes in the image obtained (which could be classified as intended deviations) from small changes (which could be classified as unintended vibration) is very important. That is why a fuzzy classifier can be used as a part of a control system. To solve this problem, the following procedure has been developed (see Fig. 10.7).

Fig. 10.6 *JVC Matsushita camcorder employing fuzzy control for image stabilisation*

Design a fuzzy controller for an image stabilisation. To compare the new shot with the previous one, any picture is divided into four parts with 30 points each and the signals received at these points are saved. Develop the rules set. Include in the set rules like:

If there are small, equally oriented deviations (= vibrations) **then** transmit the saved shot.
If there are big (= intended) or unequally oriented deviations (= movements) **then** transmit and save the new shot.

Determine the type of a fuzzy controller to be developed, its inputs and output.

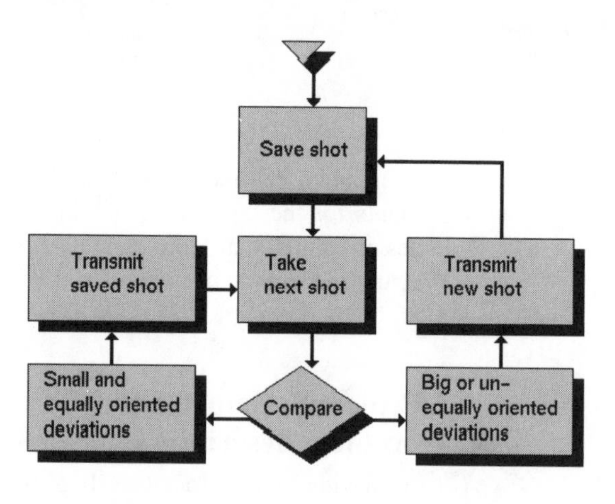

Fig. 10.7 *Procedure for image stabilisation control in a camcorder*

Project 9. Crane operational control

In an unmanned crane operational control system, the main problem is to transport the container to the target location without sway. To solve the problem several approaches have been investigated such as control methods with nonlinear feedback and the numerical methods for constrained nonlinear optimisation. However, all of them result in some performance decrease because of the sway rejection. Fuzzy control is supposed to be robust enough and can be applied here.

The mechanical model for no-sway crane operation is shown in Fig. 10.8.

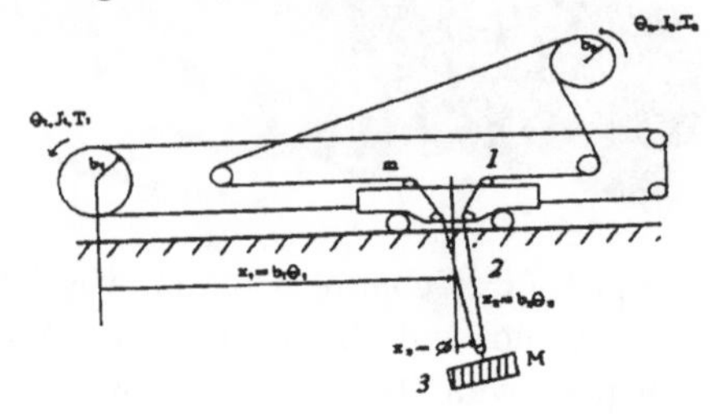

Fig. 10.8 *Schematic crane model [Baek95]*

The Euler–Lagrange motion equations can be described as:

$$\dot{x} = f(x,v,t), x = (x_1,\dots,x_6)^T, u = (v_1,v_2)^T$$

$$\dot{x}_1 = x_2$$

$$\dot{x}_2 = v_1 - \delta_1 x_3 v_2 + \delta_1 g x_3$$

$$\dot{x}_3 = x_4$$

$$\dot{x}_4 = -(v_1 - \delta_1 x_3 v_2 + (1+\delta_1) g x_3 + 2x_4 x_6) / x_5$$

$$\dot{x}_5 = x_6$$

$$\dot{x}_6 = -\delta_2 x_3 v_2 + v_2$$

where:

$$x_1 = b_1\theta_1, x_2 = \dot{x}_1, x_3 = \psi, x_4 = \dot{x}_3, x_5 = b_2\theta_2, x_6 = \dot{x}_5$$

$$v_1 = \frac{b_1 T_1}{J_1 + mb_1^2}, v_2 = \frac{b_2(T_2 + Mb_2 g)}{J_2 + Mb_2^2}, \delta_1 = \frac{Mb_1^2}{J_1 + mb_1^2}, \delta_2 = \frac{Mb_2^2}{J_2 + mb_{21}^2}$$

where J_1 and J_2 are the momentum inertia for trolley and hoist motor, M and m are the masses of the load including the spreader and the trolley including the operation room. T_1 and T_2 are the torques, b_1 and b_2 are the drum radiuses, θ_1 and θ_2 are the rotation angles for the trolley and the hoist, ψ is the sway angle and g is the gravity acceleration constant.

Design a fuzzy controller to keep the sway angle as small as possible. The sway angle is supposed to be measured.

Project 10. Control of jacks for building construction

In building and constructing, the problem of level lifting is quite typical, e.g. while restumping. In lifting, a few jacks distributed under the bottom of a construction are applied. The level of the surface should be left flat while moving the construction up or down to prevent damage. It requires the adjustment of the jack operational speeds, which is not simple especially when the construction weight is distributed unevenly around the area and several people are looking after them while they are being operated manually.

The control system (Fig. 10.9) is offered to operate the servo motors controlling the ascent and descent speeds of jacks. The construction bottom surface is equipped with level sensors S_1 and S_2. The control decision can be based on the following:

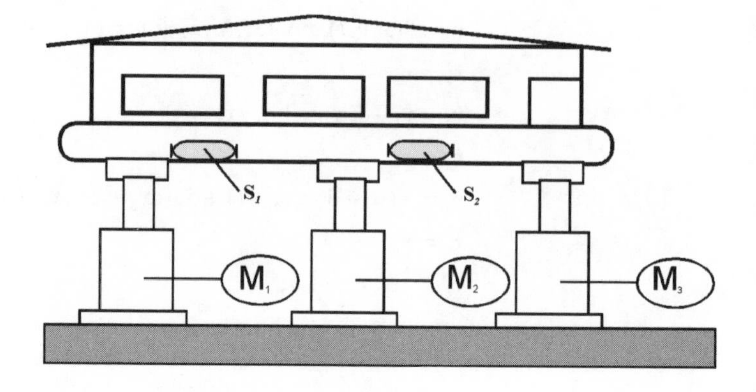

Fig. 10.9 *The control system for keeping the flat level*

If S_1 and S_2 detect the base is level, **then** speeds of all servo drivers should not change.

If S_1 detects that the left side is higher than the right, **then** the speed of M_1 should be reduced while the speeds of M_2 and M_3 remain unchanged.

If S_2 detects that the right side is higher than the left, **then** the speed of M_3 should be reduced while the speeds of M_1 and M_2 remain unchanged.

If S_1 detects that the left side is lower than the centre and S_2 detects that the right side is lower than the centre, **then** the speed of M_2 should be reduced while the speeds of M_1 and M_3 remain unchanged.

Design a fuzzy inference system (fuzzy rules) for a fuzzy controller. When does this set of rules work: in getting the construction up or down? What rules should be added to the given above? How can they be modified?

Project 11. Car movement control on an intelligent superhighway

On the future intelligent superhighway, all cars will be separated into different palettes: groups of cars travelling with the same speed. The distance between palettes as well as the distance between cars within a palette should be kept constant during travelling. In the design of the fuzzy controller, keeping a constant distance within the palette, the following car model is recommended [Dick94]:

$$m_i a_i = m_i \xi_i - K_{d_i} v_i^2 - d_{m_i} - m_i g \sin \beta_i$$

$$\dot{\xi} = -\frac{\xi_i}{\tau_i(v_i)} + \frac{u_i}{m_i \tau_i(v_i)}$$

The term $m_i \xi_i$ is the tractive engine force that the wheels apply. The variables a_i and v_i stand for the acceleration and velocity. $K_{d_i} v_i^2$ and d_{m_i} are the aerodynamic and mechanical drag forces. m_i is the total mass of the car and cargo. τ_i is the engine time lag. The term $m_i g \sin \beta$ stands for acceleration due to gravity g. The controller should be able to control the movement in up- or downhill driving when the inclination of the hill from horizontal is β_i. The control input u_i is the throttle angle of the car.

Design a fuzzy controller to control the platoon movement consisting of five cars with the parameters given in the Table 10.1. The control goal is to keep a distance deviation as small as possible. Should fuzzy controller parameters depend on the parameters of cars involved?

Table 10.1

Car parameters	Car number				
	1	2	3	4	5
Curb mass, kg	1175	1760	1020	1600	4000
Cargo mass, kg	270	200	150	400	1000
Mass total, kg	1445	1960	1170	2000	5000
τ, *sec*	0.2	0.25	0.3	0.2	0.3
K_d, *kg/m*	0.44	0.47	0.5	0.52	0.55
d_m, *N*	352	390	360	407	423

Project 12. Control of snack baking system

This project illustrates an ability of a combination of fuzzy control and expert control methods. The fuzzy control system which was developed by Omron Corp. should provide the production of high quality snacks. In this system (Fig. 10.10) some operational parameters in the oven (temperatures throughout the oven, density of the gas vapoured, colour of snacks) are measured with the sensors. The values of these parameters are specified and should be maintained while producing snacks by controlling the heaters. However, only keeping the specified parameters is not enough to produce a high quality product. The quality depends on some other parameters which cannot be measured but may be evaluated by a human expert (operator). These data includes information about cracking, tough and thickness of products. The necessity for these estimates, followed by their application in control process, requires the presence of highly qualified and experienced operators. The operators change the operational mode (heater control) of baking depending on their evaluations of the product.

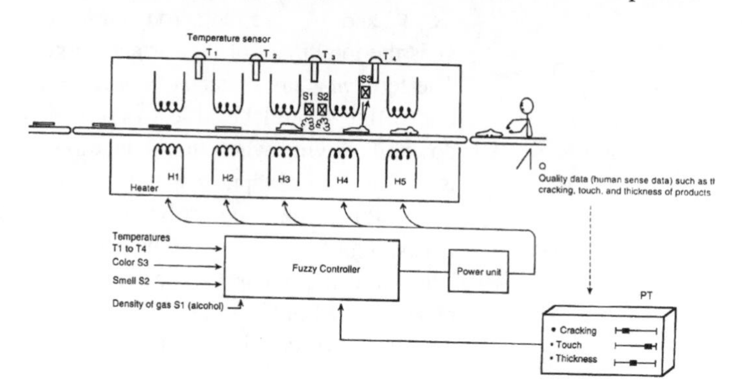

Fig. 10.10 *Fuzzy-expert control of the baking process*

The rules applied by operators in control can be formulated and added to other 'conventional' rules if a control system is designed with fuzzy methods. E.g., **if** thickness is too small **then** increase the power for heater No. 3 a little bit. The source for these rules can be the technological process knowledge.

Design a fuzzy controller providing the specified values for the product parameters measured and include some rules formulated on the base of the supposed technological process knowledge.

Project 13. Fuzzy GA controller for a refrigerator

A refrigerator with a structure that includes a fuzzy controller was developed at Samsung Electronics in 1991. This controller estimates the food temperature with the help of some fuzzy rules based on measurement results of the inside temperature, the outside temperature and the door opening/closing time. The goal of the fuzzy controller is to adjust the fan and the compressor to extend the percentage of time when the refrigerator works without consuming electricity, providing the required temperature mode. The food temperature mode is specified with fuzzy terms.

Design a fuzzy controller to solve this problem. Apply genetic algorithms (GA) to learn the fuzzy inference rules for temperature derivation based on the data supplied.

Fig. 10.11 *Fuzzy controlled fridge*

Project 14. Fuzzy GA controller for inverted pendulum control

This problem is so famous that it does not need any description. It has become a test design for fuzzy control. A fuzzy PD-like fuzzy controller is supposed to be designed with the angle error and the changing rate of it as the controller inputs. The rule table is similar to the one given in Section 3.5.3. GA are to be applied for the adjustment of the membership functions. The triangular membership functions are proposed, with any one determined by the set of three parameters (P,Q,R), see Fig. 10.12.

This set determines the shape of the membership function for each linguistic variable. They have to be adjusted with GA to improve the overall system response. Based on the experience in [Huang95] the parameters of GA could be advised as following:

Population size:	100
Number of generations:	5
Crossover rate:	0.8
Mutation rate:	0.001

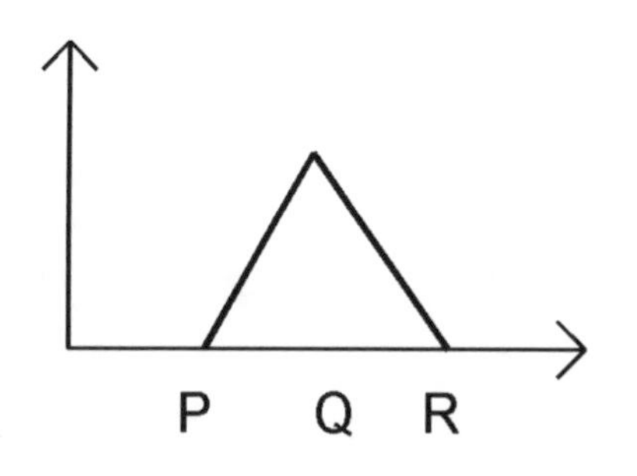

Fig. 10.12. *Three-parameter membership function*

Project 15. Flip-flop electronic circuit for fuzzy operation realisation

Fuzzy logic can be characterised as an extension of two-valued Boolean logic, where NOT, AND and OR operators are extended to fuzzy negation, *t* norm and *s* norm, respectively. In this case, the fundamental fuzzy logical circuits can be considered as one of the fuzzy extensions of a combinatorial circuit in two-valued Boolean logic. In such a situation, the concept of a fuzzy extension of a sequential circuit, which is the complex of a combinatorial circuit and memory modules in two-valued Boolean logic should be discussed. So the concept of fuzzy flip-flop appears. It is a fuzzy extension of two-valued J–K flip-flop.

Design a fuzzy logic circuit for implementing fuzzy operations like negation, *t* norm, and *s* norm based on a conventional flip-flop circuit.

GLOSSARY

Adaptive controller Controller changing the control signal and the parameters of itself depending on the changeable plant and environment model

Antecedent part of a rule An initial part of a rule which determines the conditions under which the rule is fired

Applicability degree A degree in which all *antecedent* parts of a rule correspond to it

Backpropagation method A method of recalculating the weight factors in ANN, based on propagation of the error value from the output level back to the input levels and analysis of any weight influence on the error value

Boolean logic A binary-type logic where any variable may have two values only: true or false

Characteristic function The function which determines a set having the value of unity for all elements of the set and zero for all others

Consequent part of a rule A final part of the rule which determines what will happen if the rule is fired

Control surface A plot (usually three-dimensional) which virtually presents how the controller output depends on the inputs. Frequently used to estimate the controller quality

Clipping a fuzzy set Cutting the degree of all the elements on the specified level

Crisp (conventional) set A collection of objects of any kind. In fuzzy set theory, it can be considered a specific type of the fuzzy set with a membership function having the value of unity for the elements of the crip set and zero for others

Defuzzification A process of producing a crisp output on the base of a fuzzy input. Different defuzzification methods have been developed and applied

Extension principle The rule defining the fuzzy set on the universe which has functional relationship with the initial universes. Serves as a base for the determination of the algebraic operations with fuzzy sets (see Definitions 2.11 and 2.12)

Feedback structure The structure which assumes the application of the output signal at the structure input together with the input signals. The most frequently applied in control systems, allows a comparison of the real value of the output with the specified one

Fuzzification A process of producing a fuzzy input on the base of a crisp one.

Fuzzy (logic) controller A control system which implies the methodology of fuzzy logic. This type of controller includes a knowledge-based system consisting of fuzzy rules and a special mechanism of their processing (see *Inference engine*) as its main parts

Fuzzy logic There is a narrow and a broad sense of this term. In its narrow sense, fuzzy logic is a system of logical operators defined by a calculus of interaction, that attempts to construct a model for the various modes of human reasoning, which are approximate rather than exact. In its broad sense, fuzzy logic refers to a general class of fuzzy set theories, describing a family of classes with unsharp fuzzy boundaries

Fuzzy number A fuzzified real number. See Definition 2.13

Fuzzy relation A relation of two (or in a general case *n*) fuzzy sets (see Definition 2.16). Applied in rules processing

Fuzzy rule A rule of the **If** ... **then** structure with vague predicates

Fuzzy set A set of ordered pairs, consisting of an element of the universe of discourse and the membership degree. Any element of the universe belongs to a fuzzy set with some degree only

Inference engine (mechanism) The specified process transforming fuzzy inputs into a fuzzy output by dealing with fuzzy rules, as a result of which the response corresponding to the inputs is produced. Different types of inference processing have been developed and applied

Level set (α-cut-set) The set containing all the elements belonging to the fuzzy set with some degree not lower than α (see Definition 2.4)

Linguistic hedge A modifier (e.g., very, rather) which is applied to change the membership function of the primary term

Linguistic label The name of the specified fuzzy set on the universe of discourse (e.g., short, long, medium)

Linguistic variable A variable, with a (linguistic) value that is a word or a word sentence (see Section 2.4)

Mamdani-type inference A type of fuzzy inference in which the consequent of each rule is a linguistic statement assigning a linguistic value to the variable. Supposes a defuzzification of the output produced

Membership degree A number between 0 and 1, characterising the degree to which the element of the universe belongs to the fuzzy set

Membership function A function which corresponds a real number between 0 and 1 (which is called the membership degree) to any generic element of the universe of discourse. The membership function determines a fuzzy set (see Definition 2.1)

Overshoot One of the performance indicators traditionally used in control engineering, especially while observing the controller unit step response. Calculated as the increase of the response during the transient period over its final value

PID-controller The controller with an output that consists of three parts, with the first one proportional to the input, the second one to the integral of the input, and the third one to the input derivative. This controller type is the most widely applied in industry. It can be realised as a PI-controller (when the derivative part is zero) or a PD-controller (when the integral part is zero)

Range of values associated with a fuzzy variable Several fuzzy sets would be defined within a universe of discourse, each with its own domain which overlaps with the domains of its neighbouring fuzzy sets

***s* norm** See *t* conorm

Scaling a fuzzy set Scaling a membership degree of all the elements of the fuzzy set by multiplying it by a factor less than a unity

Scaling factors Coefficients by which the fuzzy controller inputs and outputs are multiplied in order to use the normalised universe of discourse for inputs and outputs, that is the universe which coincides with the closed interval [–1,1]

Self-organising controller	Controller of which parameters are changed (tuned) without any changes in the plant model
Settling (settle) time	One of the performance indicators traditionally used in control engineering especially while observing the controller unit step response. Calculated as the time period necessary for the transient response to settle within the specified strip around its final value
Singleton function	A function which is unity at one particular point and zero everywhere else
Steady-state error	One of the performance indicators traditionally used in control engineering, especially while observing the controller unit step response. Calculated as the difference between the desired final value of the response after the transient period and the actual one
Sugeno-type inference	A type of fuzzy inference, in which the consequent part of each rule is a linear combination of the inputs. Does not involve a defuzzification process with the output being calculated as a linear combination of the consequents. Sometimes is referred to as the Takagi–Sugeno–Kang (TSK) method
Support	The crisp subset of the universe, containing all the elements belonging to the fuzzy set with some non-zero degree (see Definition 2.2)
***t* conorm (also known as *s* norm)**	A two-input function with special features (see Definition 2.10). Applied to determine the union of two fuzzy sets
***t* norm (triangular norm)**	A two-input function with special features (see Definition 2.8). Applied to determine the intersection of two fuzzy sets
Transfer function	A mathematical model which completely describes any element of the control system. Widely applied in controller analysis and synthesis

Truth value of a rule	On a single-rule level, rule strength is usually determined by the truth of antecedents of the rule. On a multiple-rule level, the rule strength of all rules bearing the consequent is determined by the rule evaluation process.
Tuning	Changing some parameters and/or structure in order to improve some performance indicators
Universe of discourse	The set which contains all the elements considered in the problem

Trademarks

Some products that are referred to in the text and their corresponding companies are given below:

AL220, INSIGHT IIe	Adaptive Logic
CubiCalc DataEngine	HyperLogic
	Management Intelligenter Technologien (MIT) GmbH
FIDE	Aptronix
fuzzyTECH, NeuroFuzzy, *fuzzy* PLC	Inform GmbH
FuzzyCalc	FuzzyWare
FlexFuzz	MentaLogic Systems
FP-3000	Omron
GIAC,SAAC,UNAC	CICS Automation
MATLAB	The MathWorks
MATRIXx, SystemBuild, RT/Fuzzy	Integrated Systems
MS-Windows, MS-Excel	Microsoft Corp.
TIL FPL, TILShell, FC110	Togai InfraLogic
WARP	SGS-Thomson

Bibliography

[Adams95] Adams G.J. and Goodwin G.C. 'A Multivariable Control Design Toolbox.' In: *Proceedings of the International Conference 'Control-95'*, Melbourne, 20–24 October 1995, Institution of Engineers, Australia, 1995, **1**, pp. 193–197.

[Adap96] Advertising and explanation notes, World Wide Web home pages, Adaptive Logic, 1996.

[Ang95] Angstenberger J. 'DataEngine – A Software Product Family for Intelligent Data Analysis.' In: *Proceedings of the Third European Congress on Intelligent Techniques and Soft Computing (EUFIT'95)*, 29 August–1 September 1995, Aachen, Germany, **3**, pp. 1892–6.

[Baek95] Baek W.-B., *et al.* 'No-sway Crane Operating Control System Development for Unmanned Container Transporting.' In: *Fuzzy Logic for the Applications to Complex Systems Proceedings of the International Joint Conference of CFSA/IFIS/ SOFT'95*, 7–9 December 1995, Taipei, Taiwan, World Scientific Publishing, 1995, pp. 489–94

[Bald87] Baldwin S.F. 'Energy Efficient Electric Motor-drive Systems.' In: *Electricity*, ed. T.B. Johansson *et al.*, Lund University Press 1987.

[Ber92] Berenji H.R. and Khedkar P. 'Learning and Tuning Fuzzy Logic Controllers Through Reinforcements.' *IEEE Transactions on Neural Networks*, **3** (5), pp. 724–40, 1992.

[Bour96] Bourmistrov A. and Reznik L. 'Hybrid Guidance Control for a Self-piloted Aircraft.' In: *Proceedings of the Conference: 'Fuzzy Logic and Management of Complexity (FLAMOC'96),'* Sydney, Australia, 15–18 January 1996, **2**, pp. 155–9.

[Chiu95] Chiu S. 'Software Tools for Fuzzy Control.' In: *Industrial Applications of Fuzzy Logic and Intelligent Systems* ed. J. Yen, R. Langari and L.A. Zadeh, IEEE Press, 1995, pp. 313–40.

[CICS96] Product Brochure, CICS Automation, Australia, 1996.

[Clel92] Cleland J. 'Fuzzy Logic Control of AC Induction Motors.' In: *Proceedings of IEEE International Conference on Fuzzy Systems*, San Diego, USA, March 1992, IEEE, 1992, pp. 843–50.

[Cox94] Cox E. *The Fuzzy Systems Handbook. A Practitioner's Guide to Building, Using, and Maintaining Fuzzy Systems*. AP Professional, Cambridge, 1994.

[Dick94] Dickerson J.A. and Kosko B. 'Ellipsoidal Learning and Fuzzy Throttle Control for Platoons of Smart Cars.' In: *Fuzzy Sets, Neural Networks, and Soft Computing*, ed. R. Yager and L.A. Zadeh Van Nostrand Reinhold, New York, 1994, pp. 63–84.

[Drian93] Driankov D., Hellendoorn H. and Reinfrank M. *An Introduction to Fuzzy Control*, Springer-Verlag Berlin, Heidelberg, 1993.

[Far92] Farrell J. and Baker W. 'Learning Control Systems.' In: *An Introduction to Intelligent and Autonomous Control*, ed. P.J. Antsaklis and K.M. Passino, Kluwer Academic, 1992, pp. 237–62.

[Fu64] Fu K. 'Learning Control Systems.' In: *Computer and Information Sciences*, ed. J. Tou and R. Wilcox, Spartan, 1964.

[Fuz92] Fuzzy Logic Educational Program, Center for Emerging Computer Technologies, Motorola Inc., 1992.

[Geb96] Gebhardt J. and von Altrock C. 'Recent Successful Fuzzy Logic Applications in Industrial Automation.' In: *Proceedings of the Fifth IEEE Conference on Fuzzy Systems*, September, New Orleans, USA, IEEE Press, 1996.

[Ghan95-1] Ghanayem O. and Reznik L. 'A New Reasoning Approach and its Application in Power System Stability.' In: *Proceedings of the Third European Congress on Intelligent Techniques and Soft Computing (EUFIT'95)*, 29 August–1 September, 1995, Aachen, Germany, **3**, pp. 1527–32.

[Ghan95-2] Ghanayem O. and Reznik L. 'A Hybrid AVR-PSS Controller Based on Fuzzy Logic Technique.' In: *Proceedings of the International Conference 'Control-95'*, Melbourne, 20–24 October 1995, Institution of Engineers, Australia, 1995, **2**, pp. 347–51.

[Guo94] Guo S. and Peters L. 'A Reconfigurable Analog Fuzzy Logic Controller.' In: *Proceedings of the Third IEEE Conference on Fuzzy Systems*, 26–29 June, Orlando, USA, IEEE Press, 1994, **1**, pp. 124–8.

[Guo95] Guo S. and Peters L. 'A High-speed Reconfigurable Fuzzy Logic Controller.' *IEEE Micro*, **15** (6), 1995, p. 65.

[Har93] Harris C.J., Moore C.G. and Brown M. *Intelligent Control: Aspects of Fuzzy Logic and Neural Nets*, World Scientific., Singapore, 1993.

[Hel94] Hellendooorn H. and Palm R. *Fuzzy System Technologies at Siemens R&D Fuzzy Sets and Systems*, **63** (3), 1994, pp. 245–69.

[Hir93] Hirota K. (ed.), *Industrial Applications of Fuzzy Technology*, Springer-Verlag Tokyo, 1993.

[Hiy89] Hiyama, T. 'Application of Rule Based Stabilising Controller to Electrical Power System.' *IEE Proceedings-C*, **136** (3), May 1989, pp. 175–81.

[Hsu93] Hsu, Y. and Cheng, C. 'A Fuzzy Controller for Generator Excitation Control.' *IEEE Transactions on Systems, Man and Cybernetics*, **23** (2), March/April 1993, pp. 532–9.

[Huang95] Huang S.-J. and Hung C.-C. 'Genetic-evolved Fuzzy Systems for Inverted Pendulum Controls.' In: *Fuzzy Logic for the Applications to Complex Systems Proceedings of the International Joint Conference of CFSA/IFIS/SOFT'95*, 7–9 December 1995, Taipei, Taiwan, World Scientific, 1995, pp. 35–40.

[Inn95] *Innovation of the Year*, EDN, 13 April 1995, p. 19.

[Ishig92] Ishigame A., Ueda T., *et al.* 'Design of Electric Power System Stabilizer Based on Fuzzy Control Theory.' *IEEE Conference on Fuzzy Systems*, 8–12 March, 1992, IEEE, 1992, pp. 973–80.

[Jam94] Jamshidi M. 'On Software and Hardware Applications of Fuzzy Logic.' In: *Fuzzy Sets, Neural Networks, and Soft Computing* ed. R Yager and L. A. Zadeh, Van Nostrand Reinhold, New York, 1994, pp. 396–430.

[Jang95] Jang J.-S.R. and Sun C.-T. 'Neuro-fuzzy Modelling and Control.' *Proceedings of the IEEE*, **83** (3), March 1995, pp. 378–404.

[Jar94] Jaramillo-Botero A. and Miyake Y. 'A High Speed Parallel Architecture for Fuzzy Inference and Fuzzy Control of Multiple Processes.' In: *Proceedings of the Third IEEE Conference on Fuzzy Systems*, 26–29 June, Orlando, USA, IEEE Press, 1994, **3**, pp. 1765–70.

[Kah95] Kahaner D.K. Fuzzy Control Applications in Japan, Asian Technology Information Program, Harks Roppongi Building, 1F, 6–15–21 Rappongi, Tokyo, Japan.

[Kim95] Kim J., *et al.* 'Industrial Applications of Intelligent Control at Samsung Electronics Co. – In the Home Appliances Division.' In: *Fuzzy Logic for the Applications to Complex Systems Proceedings of the International Joint Conference of CFSA/IFIS/SOFT'95*, 7–9 December 1995, Taipei, Taiwan, World Scientific, 1995, pp. 478–82.

[Kre92] Kreinovich V., *et al.* 'What Non-linearity to Choose? Mathematical Foundations of Fuzzy Control.' In: *Proceedings of the 1992 International Fuzzy Systems and Intelligent Control Conference*, Louisville, KY, USA, 1992, pp. 349–412.

[Krist92] Kristinsson K. and Dumont G.A. 'System Identification and Control Using Genetic Algorithms.' *IEEE Transactions on Systems, Man and Cybernetics*, **22** (5), 1992, pp. 1033–46.

[Kung93] Kung S.Y. *Digital Neural Networks*, Prentice Hall, Englewood Cliffs, 1993.

[Lea92] Lea R.N. and Chowdhury I. 'Design and Performance of the Fuzzy Tracking Controller in Software Simulation.' *IEEE Conference on Fuzzy Systems*, 8–12 March 1992, IEEE, 1992, pp. 1123–30.

[Lee90] Lee C.C. 'Fuzzy Logic in Control Systems: Fuzzy Logic Controller.' *IEEE Transactions on Systems, Man and Cybernetics*, **20** (2), 1990, pp. 404–35.

[Lin91] Lin C.-T. and Lee C.S.G. 'Neural Network Based Fuzzy Logic Control and Decision System.' *IEEE Transactions on Computers*, **40** (12), 1991, pp. 1320–36.

[Lin95] Lin C.-T. and Lu Ya-C. 'A Neural Fuzzy System with Linguistic Teaching Signals.' *IEEE Transactions on Fuzzy Systems*, **3** (2), 1995, pp. 169–89.

[Link95] Linkens D.A. and Nyongesa H.O. 'Evolutionary Learning in Fuzzy Neural Control Systems Methods for Autodesign of Systems based on Soft Computing Techniques and Human Preference.' In: *Proceedings of the Third European Congress on Intelligent Techniques and Soft Computing*, Aachen, Germany, 1995, **1**, pp. 990–5.

[Luo95] Luo J. and Lan E. 'Fuzzy Logic Controllers for Aircraft Flight Control.' In: *Fuzzy Logic and Intelligent Systems* ed. H. Li and M. Gupta, Kluwer Academic, 1995, pp. 85–124.

[Mas94] Mason D.G., Linkens D.A., *et al.* 'Automated Delivery of Muscle Relaxants using Fuzzy Logic Control.' *IEEE Engineering in Medicine and Biology*, **13** (5), 1994, pp. 678–86.

[McNeil 94] McNeil F.M. and Thro. F. Fuzzy Logic: A Practical Approach, Boston. AP Professional, 1994.

[Mik95] Miki T. and Yamakawa T. 'Fuzzy Inference on an Analog Fuzzy Chip.' *IEEE Micro*, **15** (5), October 1995.

[Nel91] Nelson M.M. and Illingworth W.T. *A Practical Guide to Neural Nets*, Addison-Wesley, 1991.

[Palm95] Palm R. 'Scaling of Fuzzy Controllers Using the Cross-correlation.' *IEEE Transactions on Fuzzy Systems*, **3** (1), 1995, pp. 116–23.

[Pav92] Pavel L. and Chelaru M. 'Neural Fuzzy Architecture for Adaptive Control.' In: *Proceedings of IEEE International Conference on Fuzzy Systems*, San Diego, USA, March 1992, IEEE, 1992, pp. 1115–22.

[Polk93] Polkinghorne M.N., Roberts G.N. and Burns R.S. 'Small Marine Vessel Application of a Fuzzy PID Autopilot.' In: *IFAC Congress Proceedings*, Sydney, 1993, **5**, pp. 409–12.

[Pres95] Presberger T. and Koch M. 'Comparison of Evolutionary Strategies and Genetic Algorithms for Optimisation of a Fuzzy Controller.' *Proceedings of the Third European Congress on Intelligent Techniques and Soft Computing*, Aachen, Germany, 1995, **3**, pp. 1850–6.

[Proc79] Procyk T.J. and Mamdani E.H. 'A Linguistic Self-organising Process Controller.' *Automatica*, **15** (1), 1979, pp. 15–30.

[Rez93] Reznik L. and Stoica A. 'Some Tricks in Fuzzy Controller Design.' In: *Proceedings of Australia and New Zealand Conference on Intelligent Information Systems, ANZIIS–93*. IEEE, Perth, Western Australia, 1993, pp. 60–4

[Rez95-1] Reznik L. and Shi J. 'Fuzzy Controller Design From a Practitioner's Point of View – Membership Function Choice.' *Australian Journal of Intelligent Information Processing Systems*, **2** (2), 1995, pp. 38–46.

[Rez95-2] Reznik L. and Little A. 'Fuzzy Controller Design From a Practitioner's Point of View – The Review of Methodologies.' *Australian Journal of Intelligent Information Processing Systems*, **2** (4), 1995, pp. 1–9

[Sanz94] Sanz A. 'Analog Implementation of Fuzzy Controller.' In: *Proceedings of the Third IEEE Conference on Fuzzy Systems*, 26–29 June, Orlando, USA, IEEE Press, 1994, **1**, pp. 279–83.

[Schw92] Schwartz D.G. and Klir G.J. 'Fuzzy Logic Flowers in Japan.' *IEEE Spectrum*, July 1992, pp. 32–5.

[Shi95] Shi J. 'Development of a Fuzzy Logic Controller as a Power System Stabiliser Using Hierarchical Structure.' In: *Industrial and Engineering Applications of Artificial Intelligence and Expert Systems. Proceedings of the Eighth International Conference*, Melbourne, Australia, 6–8 June 1995, pp. 401–9.

[Sil90] De Silva C.W. 'Fuzzy Adaptation and Control of a Class of Dynamic Systems.' *Proceedings of the Fifth IEEE International Symposium on Intelligent Control*, Philadelphia, 1990, pp. 304–9.

[Sur94] Surmann H., *et al.* 'What Kind of Hardware is Necessary for a Fuzzy Rule Based System?' In: *Proceedings of the Third IEEE Conference on Fuzzy Systems*, 26–29 June, Orlando, USA, IEEE Press, 1994, **1**, pp. 274–8.

[Tak91] Takagi H. and Hayashi I. 'NN-driven Fuzzy Reasoning.' *International Journal of Approximate Reasoning*, **5** (3), 1991, pp. 191–212.

[Tak95-1] Takagi H. 'Methods for Autodesign of Systems Based on Soft Computing Techniques and Human Preference.' In: *Proceedings of the Third European Congress on Intelligent Techniques and Soft Computing*, Aachen, Germany, 1995, **1**, pp. 646–50.

[Tak95-2] Takagi H. 'Applications of Neural Networks and Fuzzy Logic to Consumer Products.' In: *Industrial Applications of Fuzzy Logic and Intelligent Systems*, ed. J. Yen, R. Langari, L. Zadeh, IEEE Press, New York, 1995, pp. 93–105.

[Tan95] Tanaka K. 'Design of Fuzzy Controllers Based on Frequency and Transient Characteristics.' In: *Fuzzy Logic and Intelligent Systems*, ed. H. Li and M. Gupta, Kluwer Academic, 1995, pp. 187–212.

[Tang87] Tang K.L. and Mulholland R.J. 'Comparing Fuzzy Logic with Classical Controller Designs.' *IEEE Transactions on Systems, Man and Cybernetics*, 1987, **SMC-17** (6), pp. 1085–7.

[Tao94] Tao W. and Burkhardt H. 'Application of Fuzzy Logic and Neural Network to the Control of a Flame Process.' In: *Proceedings of the Second IEE Conference on Intelligent Systems Engineering*, Hamburg, September 1994, IEE, London, 1994, pp. 235–40.

[Ter94] Terano T., *et al.* (ed.) *Applied Fuzzy Systems.* Academic Press, 1994.

[Tog86] Togai M. and Watanabe H. 'Expert System on a Chip: an Engine for Real-time Approximate Reasoning.' *IEEE Expert Systems Magazine*, **1**, 1986, pp. 55–62.

[Tsyp73] Tsypkin Y. *Foundations of Learning Systems*, Academic Press, 1973.

[Ung94] Ungering A.P. and Goser K. 'Architecture of a 64-bit Fuzzy Inference Processor.' In: *Proceedings of the Third IEEE Conference on Fuzzy Systems*, 26–29 June, Orlando, USA, IEEE Press, 1994, **3**, pp. 1776–80.

[Wang94-1] Wang Li-Xin 'A Supervisory Controller for Fuzzy Control Systems that Guarantees Stability.' *IEEE Transactions on Automatic Control*, **39** (9), 1994, pp. 1845–7.

[Wang94-2] Wang Li-Xin. *Adaptive Fuzzy Systems and Control: Design and Stability Analysis*, Prentice Hall, Englewood Cliffs, 1994.

[Wu96] Wu J.C. and Liu T.S. 'A Sliding-mode Approach to Fuzzy Control Design.' *IEEE Transactions on Control System Technology*, **4** (2), 1996, pp. 141–51.

[Yag94] Yager R.R. and Filev D.P. *Essentials of Fuzzy Modeling and Control*, John Wiley, 1994.

[Yam86] Yamakawa T. and Miki T. 'The Current Mode Fuzzy Logic Integrated Circuits Fabricated by the Standard CMOS Process.' *IEEE Transactions on Computers*, **C-35** (2), 1986, pp. 161–7.

[Yas85] Yasunobu, S. *et al.* 'Automatic Train Operation System by Predictive Fuzzy Control.' In: *Industrial Applications of Fuzzy Control* ed. M. Sugeno, Elsevier (N. Holland), 1985, pp. 1–18.

[Yen92] Yen J-Y., Lin C.-S., *et al.* 'Servo Controller Design for an Optical Disk Drive using Fuzzy Control Algorithm.' *IEEE Conference on Fuzzy Systems*, 8–12 March, 1992, IEEE, pp. 989–97.

List of examples

Index

Terms marked with an asterisk are included in the Glossary (see Chapter 11)